AF726085

Teach Me AI Like I'm A CEO©

Leading & Thriving in the Age of Artificial Intelligence

My University of Oxford UK Journey in AI

VOL 1

By

L. I. Bertrand

AI Oxford | M.I.S.M. | MBA | B.A.

Dedication

My Husband Ryan

Mom Jelena

My Siblings

Oba Ghandi

Dayo B-L

***My Eternal Twin, Uju Benedicta**

Teach Me AI Like I'm A CEO©

Leading & Thriving in the Age of Artificial Intelligence

My University of Oxford UK Journey in AI

VOL 1

By

L. I. Bertrand

AI Oxford | M.I.S.M. | MBA | B.A.

Self-Published

In The United Sates of America

July 11th, 2026

ISBN: 978-1-7357665-6-0

First Edition: July 11th, 2026

Self-published in The United States of America

1830 Bertrand Drive. #62361

Lafayette LA 70596

Teach Me AI Like I'm A CEO ©

ISBN:

First Edition: July 11th, 2026

Self-published in The United States of America

1830 Bertrand Drive. #62361 Lafayette LA 70596

Contents

PREFACE

Why I Wrote This Book

It began with a dream, one I had held since my early years as a young student at the University of Ibadan. Oxford had always been more than a university in my mind. It was a very bright beacon. I used to proudly write "Affiliated with the University of Oxford, UK" on my documents in my teenage years, an aspiration and a silent promise to myself.

After decades in corporate America, on the tip of retirement and a move to Lagos, I realized there was still so much work to do in nations around the world and decided rest was not an option. My husband had always proclaimed I was "Oxford material" and so when I discovered "Artificial Intelligence for Business," was offered by the Saïd Business School at the University of Oxford, it felt like fate. I recognized AI as a major game changer by how fast it could accelerate human, organizational and national trajectories. The application process was rigorous, and when I read my admission letter, I was overwhelmed.

Stepping into that famous #1 University in the world, I knew my life journey was about to transform forever. The relationships built with my cohort were just as important as the learning. Our weeks of intense classes would be followed by weekends of assignments, and after a brief pause, we would dive into the next module. This rhythm carried on for a full year. At the end, it felt surreal. No one wanted it to end. We all wished we could rewind and do it all again.

We concluded with a grand ball, wearing tuxedos and colorful elegant gowns, followed by a superb dinner. It was a truly magical ending. By then, our cohort had formed bonds that would last a lifetime. But we still had our final exam to turn in, and after over two hundred hours of studying for the final exam, it was all over.

One of the promises I made to myself when I started studying Artificial Intelligence at the University of Oxford was that I would write a book covering my AI Oxford experience once the course was completed. I have written self-published books before, once even writing two at the same time, leaving me exhausted. So this time, I figured I would finish my last book, "God Gives Us Miracles", (now on Amazon.com), then fully focus on the book on AI.

I began putting its framework together, starting with the title that came to me right as I had settled into bed. I quickly got up, wrote it down and allowed my experiences at the University of Oxford, and teaching AI Masterclasses to script the next few chapters.

My hope is to publish Teach Me AI Like I'm A CEO, around my graduation on July 11, 2026. My twin sister was supposed to stand by me on that day. She was my greatest champion.

She passed away before I took my Oxford AI final exams, right after completing her own exams in Nursing. Her daughter had to walk across the stage to collect her certificate. I pray that I will walk across the Oxford Sheldonian auditorium, carrying her memory and completing this AI journey she had championed and celebrated.

I had no idea how much this AI Oxford journey would change my life. I did not come into Artificial Intelligence as a software engineer, a programmer, or a data scientist. I came into it as a life and leadership coach, a business executive, consultant and a lifelong learner, someone deeply curious about what was happening in the world around me. Much like the professionals and executives reading this book today.

By 2023, Artificial Intelligence was everywhere. News channels were talking about it. Business conferences were discussing it, and it was the main topic on boardroom agendas. Executives wanted clarity on it and everyday there were headlines predicting that AI would transform industries, eliminate jobs, create new jobs, change education, revolutionize healthcare, and reshape the future of humanity.

The excitement and commotion was real. I soon noticed that many professionals were struggling to separate fact from fiction. Some people thought AI was great for humanity, others thought it was dangerous. Many people simply did not understand it. I found that executives, board members, government leaders, business owners, and professionals were often expected to make decisions about Artificial Intelligence without truly understanding what it was and this concerned me. I believe people make better decisions when they have better information. As I progressed through my studies at Oxford, I started conducting AI awareness sessions for professionals, executives, women leaders, students, and organizations.

My goal was simple. I wanted people to understand enough about Artificial Intelligence to participate confidently in the conversations that were already taking place around them. I wanted them to understand what AI could do and what it could not do. Most importantly, I also wanted them to understand how AI could affect their careers, their businesses, and their futures.

This book is the result of that journey. It is not a technical manual. It is not a coding textbook. It is not written for engineers. It is written for executives, board members, business leaders, and professionals who want to understand Artificial Intelligence without needing a degree in computer science.

My approach is simple. No jargon, no complicated explanations. Just practical, understandable conversations about one of the most important technologies of our lifetime. If you have ever sat in a meeting and heard people discussing Artificial Intelligence while secretly wondering what they were talking about, or just wish to develop your AI awareness and use, this book is for you. It is part personal AI experience story, practical executive guidance, and AI education.

There are many technical AI books, but there are very few books written by someone who went from United States, Abidjan, Cote D'Ivoire, Guinea, Conakry, Lagos, Nigeria, Windhoek, Namibia corporate experience, to Oxford UK AI studies, multiple international AI summits, teaching thousands of professionals how to understand AI. That perspective is what makes this book unique.

INTRODUCTION

My Journey to Oxford and Artificial Intelligence

As a graduate of the University of Ibadan, I was aware of the historical connections between the two institutions, and I always viewed Oxford with a sense of respect and admiration. Life, however, had other plans. My career took me in different directions. I spent decades building a professional life, working across different sectors and countries. Along the way, I earned three Masters degrees in Information Systems, specializing in Organizational Leadership, Business Administration, and Marketing. By the time I returned to Lagos with its year round summer, and after many years in the corporate world, I thought I was entering a season of slowing down. Instead, I found myself entering one of the most exciting and active chapters of my life.

I quickly realized there was still far too much work to do. There were communities to serve, organizations to support and people to help navigate through life. Retirement would have to wait. At around the same time, Artificial Intelligence was beginning to dominate conversations around the world.

Everywhere I turned, people were talking about AI. Some people were excited. Some were nervous. Most were confused. I became curious. The more I listened and learned, the more I realized that Artificial Intelligence was not simply another technology trend. Something much bigger was happening. The world was evolving. The workforce as we knew it was changing. Leadership focus was shifting. I knew I wanted to understand it properly. Not from social media or from headlines or secondhand opinions. I wanted to learn from the #1

University in the world, University of Oxford. My husband had often encouraged me to consider Oxford. He believed I belonged there just as I believed many years before I even met him.

As I explored my options, I discovered a program at University of Oxford's Saïd Business School on Artificial Intelligence for Business. The moment I saw it, everything clicked. This was exactly what I had been looking for. I submitted my application and waited. The process was rigorous. The standards were high.

When the acceptance letter arrived, I sat quietly for a moment and simply stared at it. I had been admitted to the University of Oxford. My assigned college was Jesus College. I was not sure how that happened, but it could only be God. Sometimes life gives us moments that feel too big to fully process. This was one of those moments. I still remember walking into Saïd Business School for the first time. Before I approached the counter, I sat down on one of the brown leather couches in the lobby. Oh yes, everything about University of Oxford is top notch. Most of the local taxis are Mercedes Benz.

For a few moments, I simply looked around. I was finally here, on the campus of The University of Oxford. The #1 Ranking University in the world. An institution with so much history and pedigree. This was a big deal. I collected my registration materials, received my identification badge, and completed all the formalities. This University of Oxford I had admired for many years was no longer an idea. It was real. My new career and life journey had just begun.

Let me invite you on this book journey with me as we delve into AI. And much like my Oxford journey, it is not just about the serious and technical AI stuff, it is also about building relationships and prioritizing the humanity in the Technology loop. I hope to see you in one of my AI Masterclasses soon.

Chapter 1

MEETING MY OXFORD AI COHORT

Over 28 Countries, One Classroom

One of the first things that struck me about Oxford was not just the amazing historic buildings, the history, or the prestige. It was the people. As we began introducing ourselves, I quickly realized that this would be unlike any other educational experience I had ever had from anywhere in the world.

A Truly Global Classroom

Our cohort represented approximately 28 countries. Some participants had traveled more than 20 hours to attend the program, coming from The United States, Australia,

Algeria, France, China, Czech Republic, Bangladesh, New Zealand, Armenia, India, Thailand, Pakistan, The United Arab Emirates, Greece, Iran, Switzerland, South Africa, Nigeria, Kenya, Mauritius, Hungary, Ireland, Italy, the United Kingdom. Our class was a mini-United Nations.

Members of the cohort brought distinct perspectives, professional backgrounds, and distinctive life experiences. Top executives, entrepreneurs, and specialists in multinational corporations and unique fields. Yet despite our differences, we had a shared conviction, and that was that the world was dramatically changing and Artificial Intelligence was playing a major role in that transformation. We wanted to understand it and be part of it.

The Energy in the Room

I remember sitting in class, looking around the room, and sensing a level of curiosity that was difficult to describe. People were asking questions, challenging ideas, sharing experiences, and connecting concepts across industries and cultures. The energy was extraordinary. You could feel the spark in the room, an excitement about learning that reminded me why education and especially at this University was so powerful.

Learning at the Speed of Oxford

Then the professors arrived, and I was not prepared for what happened next. The discussions moved fast, very fast. They were not reading from slides or textbooks. They were teaching from years of research, experience, and expertise. Ideas flowed rapidly, concepts connected effortlessly, and questions generated even deeper discussions. At times, it felt as though their minds were moving at lightning speed.

Before long, my cohort and I were speaking with the same speed and confidence. The conversations became sharper, the questions deeper, and the thinking, faster. The entire cohort seemed to rise together as every module stretched us intellectually and challenged all of our individual and collective assumptions. Every assignment compelled us to think differently. That was one of the greatest gifts of Oxford. It did not just teach us information, it taught us how to think, and it happened with rapid fire velocity, as we got immersed in the Oxford experience. The Oxford excellence is not a myth. It is real.

Learning Beyond the Classroom

This experience was not limited to the classroom. Memorable learning was also happening past class. After long hours of lectures, we often gathered for drinks, dinner, dessert, or all three, where the class discussions continued. I remember vividly thoroughly enjoying the conversations, the friendships, and the opportunity to learn from remarkable people, some of whose life experiences were very different from mine.

From Classmates to Community

Those evenings became an important part of the journey. We shared stories about our countries, talked about our careers and families, debated the future of technology, and laughed often. Slowly, strangers became friends. Over time, those friendships deepened. The cohort became more than a group of classmates. We became a community. Looking back, I can honestly say that the relationships were just as valuable as the education.

Artificial Intelligence brought us together, and friendship keeps us connected. These relationships continued long after the lectures ended. We continue to re-convene in Oxford every year with well-curated AI specialized summits. When one of my cohort, Hayk, organized a technology event in Armenia, he invited me as a special guest along with some of our Oxford Cohort to speak on AI on their international stage. My fellow cohort, Johannes and John, collaborated with me to teach AI awareness and we stay connected with our amazing University of Oxford heads of AI for Business program, Dr. Felipe and Dr Alex, our extraordinary recruitment and Administration staff, Joe and Joanna who started the journey and have stayed connected with us. It has been Truly Remarkable.

Chapter 2

Learning to Think Differently

One of the misconceptions people have about Artificial Intelligence is that it is primarily about technology. My experience has taught me that AI is also about learning to think differently and although the technology, data and algorithms matter, before any of those things can make sense, we must learn a new way of looking at problems, opportunities, and decisions.

The Oxford Rhythm

That realization began during my first few modules at Oxford. The structure of the program was rigorous. We would spend an intensive week together learning, discussing, debating, and absorbing enormous amounts of information. Then we would return home and that was when the real work began. There were moments when I read the same paragraph multiple times, stop and rethink entire concepts to truly get an understanding.

When the Mystery Began to Make Sense

Then something interesting happened. The more I worked through the assignments, the more I began to see patterns. AI was mathematics. It was statistics. It was data. It was computing power, but most importantly, it was human creativity. This realization changed everything for me once I began to understand the underlying principles.

The Leadership Gap

The more I learn about AI, the more I realize that many professionals are uneasy with this new technology they do not fully understand. This is human nature. We are often uncomfortable with what feels unfamiliar. The challenge is that AI is becoming impossible to ignore. Not only is it in boardrooms, government, healthcare, energy, education, creatives, but it is also in media, manufacturing, retail, banking, agriculture, and virtually every sector of society. As a result, many leaders are suddenly being asked to make decisions about AI without understanding even the basic concepts.

Beyond Hype

I started paying close attention to this gap. Whenever I attend meetings, conferences, or professional events, I notice that AI discussions often produce two reactions. Some people become overly excited and believe AI can solve almost every problem. Others become nervous and worry that AI will eliminate all jobs, industries, and eventually human value.

Neither position reflect current reality. The truth is somewhere in the middle. Artificial Intelligence is an incredibly powerful tool. Sophisticated, transformative, and important, but a tool, nonetheless. The real question is whether individuals, organizations, and leaders will learn how to use AI effectively.

AI Literacy as a Leadership Skill

As I moved through the Oxford program, I became increasingly convinced that AI education would become one of the most important leadership requirements of the future. Not because every executive needs to become a programmer, or a data scientist, but because every leader needs to understand enough about AI to make informed decisions.

From Curiosity to Mission

That realization planted a seed. What began as personal curiosity slowly evolved into something much larger. I found myself talking about AI with friends, then colleagues, executives, organizations, and groups of leaders. The more conversations I had, the more I realized that people were looking for practical explanations. They did not want complicated technical language. They wanted someone to explain AI in plain English, remove the mystery, and connect the technology to their everyday lives.

Without realizing it, I was beginning a second career. Many years ago, in my mind, my ideal career was one where I used my brain, and where I did not need more than my digital resources to impact a room of people. I had entered Oxford to study Artificial Intelligence, and I was leaving Oxford with a mission to teach it. And that mission continues today with AI Masterclasses that I curate from basic to advanced classes. In these classes, I break down AI concepts into bite sized chunks for ease of understanding and adoption. This book follows the same format.

Chapter 3

What Exactly Is Artificial Intelligence?

A Simple Definition

The core idea is surprisingly simple. Artificial Intelligence is when a machine performs tasks that require human intelligence which might include language translation, image recognition, predicting outcomes, providing reasonable answers to complex questions, supporting decisions, or learning from experience through data.

AI Is Already in Everyday Life

Most Artificial Intelligence today is near invisible. We use it every day, often without thinking about it, like when the phone finishes our sentence before we type it or with smart door unlock using facial recognition. This is all powered by Artificial Intelligence. AI is already woven into our daily lives. The question is not whether you use AI. The better question is whether you realize you are using it. This is why my AI Awareness Masterclasses and this book are so important, because a lot of people need awareness on the advancements in AI and how their lives or livelihoods are going to become directly or indirectly affected by it.

Why AI Feels New

One of the biggest misconceptions about AI is that it suddenly appeared in the last few years. Artificial Intelligence has been around for decades. What changed is that three powerful forces came together at roughly the same time.

1. Enormous amounts of data became available.
2. Computers became more powerful.

3. Researchers began to develop better algorithms.

AI Learns Through Patterns

Think about teaching a child to recognize candy. You show the child hundreds of pictures, then thousands. Eventually, the child learns what candy looks like. I remember when my niece was a toddler and her Mom had never given her candy. I came to visit for the first time and had no inkling and I gave her a lollipop, and she had no idea what it was. She used it to play drums on my face, then on her Mom, then on her face and suddenly, it touched her lips and that was it for that lollipop. She was exhibiting what in AI is knowns as "Unsupervised Learning".

Artificial Intelligence works in a similar way, except that machines learn through data rather than life experience. The more relevant data a system has, the better it becomes at recognizing patterns. AI identifies relationships using data, trends, similarities, and uses them to predict probabilities.

Many human activities involve recognizing patterns. The difference is that AI performs pattern recognition at a scale and speed that human beings cannot match. That does not mean AI is smarter than humans. It means AI is better suited for specific types of tasks.

What Remains Uniquely Human

In 1950, Alan Mathison Turing, an English Mathematician and Computer Scientist published a groundbreaking paper called *Computing Machinery and Intelligence*. In it he asks a practical question, "could a machine behave so intelligently that a human could not tell the difference?" If in a test the machine could not be distinguished from the human, the machine is said to have passed the Turing test. (Turing, 1950).

Humans still possess qualities that machines do not, including judgment, wisdom, and real empathy. These distinctions are not minimal. They are the qualities that make human leadership essential in the age of Artificial Intelligence.

The Real Opportunity

The future is not about humans competing against Artificial Intelligence. The future is about humans working alongside it. The professionals who thrive in the coming years will not necessarily be the ones who know the most about technology but the ones who understand how to combine human intelligence and emotional intelligence with artificial intelligence.

AI Apprehension

Throughout the many AI awareness sessions I have delivered, I have discovered something important. Most people are more concerned about not understanding it. Once people understand what AI is, how it works, and where it can be useful, much of the anxiety begins to disappear. That is exactly why I wrote this book. My goal is not to turn readers into engineers. My goal is to provide guidance where you become comfortable enough with Artificial Intelligence to participate confidently in the conversations and decisions shaping our future. I want you to use and master available AI tools. Artificial Intelligence is no longer a future technology, and it is here to stay.

Chapter 4

AI and the Trojan Horse

The human fascination with creating intelligence is hundreds of years old. Long before computers existed, people were already imagining machines that could learn, and act independently. In many ways, Artificial Intelligence began as a dream long before it became a technology. To understand where AI is going, we first need to understand where it came from. Human beings have always been fascinated by intelligence. We admire it, study it, try to improve it, and have long wondered whether it could somehow be recreated outside the human mind. Ancient civilizations told stories about objects that appeared to think for themselves. Many of these stories were myths, legends, and folklore, but they reveal something important and that is that people have been imagining intelligent machines for centuries. The dream came long before the technology.

Strategy, Behavior, and the Trojan Horse

One of the clear examples comes from the story of the wooden Trojan Horse used by the Greeks to hide their soldiers and penetrate enemy territory in the city of Troy in order to win the Trojan war in the 12th or 13th Century. The Trojan Horse was not Artificial Intelligence. It was not a machine or a computer, but rather, a military strategy. This illustrates an important concept which is that intelligence is not only about strength. It is also about planning, prediction, and understanding how people think. The horse appeared harmless, the Trojans accepted it, and the hidden soldiers inside changed the outcome of the war. In a book about AI, the story matters because modern Artificial Intelligence also relies on patterns, behavior, prediction, and decision-making.

The Trojan Horse reminds us that intelligence is not simply about force. It is about strategy, understanding how information influences decisions, and recognizing behavior. Many ideas we now associate with AI have roots in humanity's long history of studying how people think and act.

The First Mechanical Thinkers

As civilizations advanced, inventors began creating mechanical devices that could perform tasks automatically. Ancient Greek engineers built automated machines powered by water and air pressure. Centuries later, European inventors created clocks and mechanical devices that could perform predetermined actions.

These machines were not intelligent. They could not learn or reason. But they represented something important, that machines could perform tasks that had once required human effort. That was a crucial step toward modern AI.

Chapter 5

The AI Winters - When the Dream Nearly Died

Artificial Intelligence had more challenges than anyone expected, and that realization led to one of the most important periods in AI history, now coined as the AI Winter. The early pioneers of AI achieved great breakthroughs. Computers were beginning to solve mathematical problems, play simple games, prove certain logical theorems, and perform tasks that had once seemed impossible for machines.

The media became excited. Governments became interested. Universities received funding, researchers received grants, and the future looked bright. The problem was that these early successes created expectations technology could not yet fulfill. Researchers knew they were making progress, but the public believed intelligent machines were right around the corner.

The First AI Winter

By the early 1970s, enthusiasm began to cool. Funding agencies started asking difficult questions such as, where were the intelligent machines that had been promised? Why was progress taking so long? Why were costs increasing while results remained limited? Governments began reducing support. Research budgets were cut, projects were canceled, and many AI laboratories struggled to survive.

The excitement of the 1950s and 1960s gave way to skepticism, and this period became known as the First AI Winter. The term "winter" was appropriate. The field did not disappear, but growth slowed dramatically.

Investment dried up, public interest declined, and many people concluded that Artificial Intelligence had been overhyped.

The Return of Hope

Researchers continued working. New ideas emerged. Computers became more powerful, data became more available, and progress slowly resumed. By the 1980s, a new approach called Expert Systems began generating excitement. These systems attempted to capture the knowledge of human experts and translate that knowledge into computer rules. For example, a medical expert system might contain hundreds or thousands of rules to help identify possible illnesses. Businesses, investors, and governments became interested again. AI appeared to be making a come-back.

The Second AI Winter

The excitement surrounding Expert Systems eventually encountered the same challenge as earlier AI efforts and this was that Expert Systems worked reasonably well in narrow situations, but they were expensive to maintain, difficult to update, hard to scale, and lacked the flexibility of human thinking. Once again, expectations exceeded results. Funding declined, enthusiasm faded, and by the late 1980s and into the 1990s, Artificial Intelligence entered what became known as the Second AI Winter. Many organizations stopped investing. Many researchers changed fields. Many executives concluded that AI was simply not ready. For a second time, the dream appeared to be fading.

What Leaders Can Learn from AI Winters

Whenever I teach executives about Artificial Intelligence, I tell them that this chapter may be more important than the technical chapters. The story of AI is not only a technology story. It is also a leadership story, a business story, and a human story. Every major innovation experiences cycles of enthusiasm and disappointment. The lesson is not to ignore innovation or blindly believe every prediction, but to develop informed optimism. The best leaders learn how to separate long-term potential from short-term excitement. They understand that transformational technologies often take longer than expected, but when they finally mature, they can change everything.

Why AI Survived

The question has become, If AI experienced two major winters, why and how did it survive? The answer is that researchers were not wrong about the potential of Artificial Intelligence. They were simply early. The vision was right, but the technology was not ready. At that early stage, computing power was limited and data was scarce. Algorithms were not ready for their full potential. The dream had essentially arrived before the tools needed to achieve it.

Chapter 6

The AI Renaissance - How Artificial Intelligence Returned More Fierce

This chapter marks the point where the AI story moves from survival to momentum. The winters were ending, and the seeds planted by Alan Turing, John McCarthy, Geoffrey Hinton, and countless other researchers were beginning to bear fruit. By the late 1990s, many people had stopped paying close attention to Artificial Intelligence. Some believed the technology had failed. Others thought it had been oversold. Many executives had moved on to other priorities. But researchers continued working quietly, just like they are doing in 2026. Not with headlines or fanfare, but with years of steady work that most people never notice.

Google is quietly expanding data centers to accommodate this exploding technology.

The progression has continued in the same trajectory. It is looking like the hype is dying down, but do not be misguided, Artificial Intelligence is not slowing down. There are years of steady work that most people never notice, but once the results are revealed, whole swats of human jobs might be gone. Just gone. Never to return. This is why I have been running AI Masterclasses and now staying up late writing this book, so humans can learn to reskill and upskill and figure out work and means of living that are assured avenues aligned with AI.

Three Things Changed Everything

Throughout my Oxford studies, one lesson came up repeatedly and it was that major occurrence of technology breakthroughs rarely happen because of one factor. They happen when several important pieces finally come together at the same time. That is exactly what happened with AI.

Three forces changed everything.

More powerful computing

Massive amounts of data

Better algorithms

Think of AI as a high-performance sports car. For decades, researchers had many pieces of the vehicle, but they did not yet have the engine, the fuel, or the roads. Eventually, they got all three, and suddenly the car could move fast.

How the Internet Changed the Game

When the internet exploded in popularity, something quite extraordinary happened, human beings began creating enormous amounts of digital information. Every email, website, search, online purchase, social media post, digital photo, video, and click added to a growing ocean of data. This mattered because Artificial Intelligence learns from data. The more data available, the better AI systems becomes at identifying patterns. Without realizing it, billions of people were helping generate the raw material needed for modern AI.

Computers Became Incredibly Powerful

The second breakthrough was computing power. For years, researchers had ideas that computers simply could not process efficiently. The machines were too slow, the calculations were too expensive, and the hardware was too limited. Then technology advanced. Processors became faster, storage became cheaper, and cloud computing emerged. Organizations could access computing power that previously would have been impossible to afford. Problems that once took weeks could be solved in hours and problems that once took hours could be solved in minutes. The impossible started becoming possible.

The Rise of Machine Learning

The third breakthrough was perhaps the most important as researchers shifted their thinking. Instead of programming computers with endless rules, they began teaching computers to learn from data. This became known as Machine Learning, and it changed everything. Think about teaching a child to recognize a dog. You could try writing down every rule. It has four legs, a tail, fur, a certain size, a certain shape, but eventually you encounter a dog that breaks one of those rules.

A Machine Learning system works differently. Instead of memorizing rules, it studies thousands or millions of examples. Over time, it learns patterns. The machine is not thinking like a human, rather, it is identifying relationships within data. That shift is the primary game changer with Artificial Intelligence.

The Return of Neural Networks

Interestingly, one of the technologies that helped revive AI had existed for decades, and this is Neural Networks. Researchers had experimented with Neural Networks years ago, using a simplified model inspired by how neurons in the human brain connect and transmit information. For years, Neural Networks struggled because computers lacked sufficient power.

Once computing power improved, they became practical. Researchers discovered that larger networks trained on larger datasets could produce dramatically better results. The results were compelling. Machines became better at recognizing images, understanding speech, translating languages, and identifying patterns. A new era had begun.

Geoffrey Hinton and Deep Learning

One of the key figures in this story is Geoffrey Hinton, often referred to as one of the godfathers of Artificial Intelligence. While many researchers moved away from neural networks, Hinton continued to believe in their potential. For years, his ideas received limited attention. Then the technology finally caught up. His work helped fuel what became known as Deep Learning, which would become one of the foundations of modern AI. Today, Deep Learning powers many technologies we use every day, including:

Voice assistants, Image recognition, Language translation, Recommendation systems and Generative AI. Much of modern AI traces its roots back to these breakthroughs.

A Computer Defeats a World Champion

Then came one of the moments that captured the world's attention. In 1997, IBM's Deep Blue defeated world chess champion Garry Kasparov. For decades, chess had been viewed as one of the ultimate tests of human intelligence, and many people believed machines would never defeat the best human players. (Hsu, 2002).

The victory made headlines around the world. Some people celebrated, others worried, but everyone paid attention. The message was clear. Artificial Intelligence was becoming capable of tasks once thought impossible.

Watson Enters the Conversation

A few years later, IBM introduced Watson, which gained global attention by competing on the quiz show *Jeopardy!* Unlike chess, *Jeopardy!* required understanding language, interpreting clues, and retrieving information. Watson won, and once again the world took notice. Artificial Intelligence was no longer confined to laboratories, it was entering the mainstream.

The Moment Everything Changed

Despite those impressive achievements, nothing prepared the world for what happened next. In late 2022, millions of people suddenly experienced Artificial Intelligence firsthand, not through a research paper, conference, or technology demonstration, but through conversation. People typed questions into ChatGPT and received responses that felt remarkably human. For many, this was their first meaningful interaction with AI. Overnight, Artificial Intelligence became a dinner-table conversation. Boardrooms asked questions, governments paid attention, schools debated policies, and organizations began developing strategies.

Why This Matters

As remarkable as today's AI systems may seem, they did not appear overnight. They are the result of decades of research, failures, setbacks, and persistence. That is one reason I find the history of AI so inspiring. The people who built today's AI systems continued working even when funding disappeared, critics dismissed them, and progress seemed painfully slow. Innovation often works that way. The breakthroughs we celebrate publicly are usually built on years of quiet effort that nobody sees.

And as impressive as today's AI may be, I believe we are still in the early chapters of the story. The future chapters have not yet been written. In many ways, we are all living through the beginning of a new technological era. That brings us to the next question, which is, "how does AI actually learn?" To answer that, we need to understand Machine Learning.

Chapter 7

How Machines Learn Without Being Explicitly Programmed

If there is one phrase you will hear repeatedly in the world of Artificial Intelligence, it is Machine Learning. People use the term so often that many assume everyone understands it, but the truth is that many people do not. By the end of this chapter, you will. Let me begin with the simplest explanation. Machine Learning is a method of teaching computers to learn from data instead of programming every rule manually. That is the basic idea.

How We Used to Program Computers

For many years, computers operated by following very specific instructions. A programmer had to tell the computer exactly what to do. Think "Hello World". Everything had to be written into the program. Imagine teaching a computer how to approve a loan. You might create rules such as, if income is above a certain amount, approve the application. If income is below a certain amount, reject it. If debt is too high, reject it. If the credit score is strong, approve it.

In traditional software, the computer simply follows the rules. This works very well when the rules are clear. But what happens when a problem involves thousands—or even millions—of factors? That is where Machine Learning comes in.

Teaching Through Examples

Think about how a child learns. You do not hand a three year old a rulebook containing thousands of pages. You show the child examples. You point to a dog and say, "That is a dog." Then you show another dog, and another, and another. Eventually, the child recognizes dogs independently. Nobody gives the child a mathematical formula. The child learns from examples. Machine Learning works in a similar way. Instead of giving a computer every possible rule, we give it examples, lots of examples, and the machine studies those examples to identify patterns. Over time, the system becomes better at making predictions. But one distinction is important. Artificial Intelligence does not process like humans. It identifies patterns within data.

My Favorite CEO Example

When I explain Machine Learning to executives, I often use a business example. Imagine your organization has 10 years of customer data. Some customers remained loyal, while others left. A Machine Learning system can study those historical records and identify patterns that may help predict future customer behavior. For example, the system might discover that customers who contacted customer service multiple times were more likely to leave. Customers who purchased certain products tended to stay longer. Customers who used specific services often generated higher revenue. The machine learns from historical examples and uses those patterns to predict future outcomes. That is Machine Learning in action. It is not guessing, and it is not using intuition. It is identifying statistical patterns.

Supervised Learning

The first major category of Machine Learning is Supervised Learning. Think of it as learning with an answer key. A teacher gives students questions along with the correct answers, and over time, the students learn how to answer similar questions on their own. In Supervised Learning, the machine receives data along with the correct answers. For example: Customer records labeled "Stayed" or "Left." Medical images labeled "Healthy" or "Disease Detected." Emails labeled "Spam" or "Not Spam." The machine studies both the data and the answers. Eventually, it learns to make predictions on new information. Many business applications of AI rely heavily on Supervised Learning.

Unsupervised Learning

In Unsupervised Learning, the machine receives data without labels, categories, or correct answers. Its job is to identify patterns on its own. Imagine walking into a room filled with thousands of people and being asked to group them. You might organize them by age, profession, interests, or location. In a similar way, Unsupervised Learning looks for hidden relationships inside data. Businesses use this approach frequently. Retailers identify customer segments, banks detect unusual behavior, and marketing teams discover purchasing patterns. The machine helps uncover insights humans may not immediately notice.

Reinforcement Learning

The third major category is Reinforcement Learning, which is learning through trial and error. Think about teaching a child to ride a bicycle. The child tries, falls, adjusts, tries again, and gradually improves through feedback. Machine Learning can work the same way. The machine takes an action, receives feedback, and adjusts. Positive results are rewarded while negative results are discouraged. Gradually over time performance improves.

This is how many advanced AI systems learn to play games, optimize processes, and make decisions. The machine is not told exactly what to do. It learns through experience.

Why Machine Learning Matters

Why has Machine Learning become such a big deal? The answer is because the world now generates more data than humans can manually analyze because every second organizations create enormous amounts of information, including

1) Customer transactions
2) Medical records
3) Sensor readings
4) Financial data
5) Supply chain information
6) Security logs
7) Social media activity

Machine Learning allows organizations to find value in data. It helps identify hidden patterns, support better decisions, improve efficiency, and predict future outcomes.

What Machine Learning Cannot Do

Machine Learning is powerful and can only learn from the information it is given. If the data is poor, the results will often be poor. If the training data contains bias, the outcomes may contain bias. If the data is incomplete, the predictions may be inaccurate. That is why data quality matters so much. The machine may be sophisticated, but it is still learning from the information we provide. That brings us to the next question. If Machine Learning learns from data, how does it actually process that information? To answer that, we need to explore one of the most important concepts in modern AI. Neural Networks.

Chapter 8

Artificial Neural Networks Explained

Why Researchers Looked at the Human Brain

One of the most fascinating things about Artificial Intelligence is that many of its ideas were inspired by nature, especially the human brain. A neural network does not think, feel, or become conscious the way a human being does. (Although with some AI programs in 2026, I see 'thinking' on the screen as it is processing information. This can be misleading). What inspired researchers was one important observation and that was that the brain learns through connections with billions of neurons and communicate with one another. Information flows through these connections, patterns are recognized, and decisions are made. Researchers wondered whether computers could learn in a similar way. That question led to the development of artificial neural networks.

An Executive Analogy

Imagine you are the CEO of a large company. Every day, information flows through your organization such as customer complaints, sales reports, financial reports, market research, and operational data. You do not personally review every detail. Instead, information moves through layers of the organization. Managers review it, directors refine it, vice presidents interpret it, and eventually a recommendation reaches the executive team.

A neural network works in a somewhat similar way. Information enters the system, multiple layers process it, patterns are identified, and a conclusion is produced. The process is mathematical rather than organizational, but the basic idea is that information becomes more useful as it moves through layers of processing.

Three Basic Layers

Most neural networks are built around three main components

1) Input layer receives the information entering the system.
2) Hidden layers process the information and identify patterns.
3) Output layer produces the final result or prediction.

For example, imagine we are teaching a system to recognize whether a picture contains a cat. The image enters through the input layer. The hidden layers analyze shapes, edges, colors, textures, and relationships. Finally, the output layer produces a result along the lines of whether or not it is a cat.

Why They Are Called Hidden Layers

One question I am often asked is why these layers are hidden. We can see the information going into the network, and we can see the result coming out, but what happens in between can be difficult to interpret. Inside those layers, the network develops some internal relationships and calculations. Even the engineers who build these systems cannot always explain every decision being made inside the model. Can you believe this? Is it scary? Maybe. This is sometimes called the "black box" problem, where the answer may be accurate, but understanding exactly how the network arrived at that answer can be shrouded in mystery.

Learning Through Adjustment

The most interesting part is that neural networks improve over time. Imagine a student taking a practice exam. The student gets some answers wrong, receives feedback, adjusts, and tries again. Eventually, performance improves. Neural networks work in a similar way. The network makes a prediction, compares that prediction with the correct answer, adjusts its internal calculations, and repeats the process, sometimes millions of times. Over time, it becomes better at recognizing patterns. The machine is not learning the way humans learn, but it is learning, nonetheless.

Why More Data Matters

One reason modern AI has advanced so rapidly is that today's neural networks have access to enormous amounts of data. Think about how children learn. A child who sees ten pictures of dogs learns something, but a child who sees ten thousand pictures learns much more. The same principle applies to AI. More relevant examples generally produce better results. This is one reason companies place such a high value on data. In many ways, data is the fuel that powers Artificial Intelligence; without it, even the most sophisticated neural network is useless.

Deep Learning: A Powerful Subset of AI

As researchers added more layers to neural networks, performance improved dramatically. These larger systems became known as deep neural networks, and the field became known as Deep Learning. Deep Learning and Artificial Intelligence are often used interchangeably, but they are not the same thing.

Deep Learning is a subset of AI. Think of Artificial Intelligence as the entire automobile industry. Deep Learning is one powerful type of vehicle within that larger ecosystem.

Why Neural Networks Matter

Neural networks matter because many of today's most impressive AI achievements rely on them. They power or support technologies such as

- Facial and speech recognition
- Language translation
- virtual assistants
- Medical imaging
- Fraud detection

- Self-driving vehicle research
- Generative AI
- ChatGPT
- Image generation tools
- Recommendation systems

What Neural Networks Cannot Do

Neural networks are extraordinary, but they do not possess wisdom, judgment, ethics, common sense, or an understanding of consequences the way humans do. A neural network can recognize patterns. A human being can understand meaning. That distinction matters, and it is one reason human leadership remains critically important in the age of AI. Technology can help us make decisions, but it should not replace our responsibility for those decisions.

The Next Big Leap

Neural networks gave Artificial Intelligence the ability to recognize patterns at an entirely new level. But another breakthrough was coming. Researchers began asking a new question, what if machines could understand and generate human language, not just individual words, or sentences, but entire conversations?

That question would eventually lead to one of the most transformative developments in modern technology, Large Language Models.

Chapter 9

Large Language Models and Generative AI Explained Like You're a CEO

When AI suddenly became everyone's favorite topic, People were suddenly talking about it in boardrooms, governments, universities, the media. It seemed as though the entire world had discovered Artificial Intelligence overnight. Of course, AI had been around for decades. What changed was that people could finally interact with it directly. For many, that first meaningful interaction came through ChatGPT. Suddenly, AI was no longer something hidden in a research laboratory. It was on phones and laptops. You could ask questions, hold a conversation, request a summary, brainstorm ideas, or ask it to write, and it would respond in seconds.

OpenAI introduced ChatGPT on November 30, 2022, describing it as "a model called ChatGPT which interacts in a conversational way" and noting that the dialogue format allows it to answer follow-up questions, admit mistakes, challenge incorrect premises, and reject inappropriate requests. (OpenAI, 2022). There was no moat and it was all aboard the AI train. Suddenly the playing field was level and everyone had access.

What Is a Large Language Model?

A Large Language Model, often called an LLM, is an AI system trained on enormous amounts of data so it can recognize patterns in human language and generate responses. Think about all the books, articles, websites, reports, research papers, and documents that exist in digital form. If you studied vast amounts of that material, you would begin to notice which words commonly appear together, how sentences are structured, and how ideas relate to one another. It studies patterns in language at an extraordinary scale.

Why It Feels Human

Large Language Models do not possess consciousness, emotions, life experiences, or innate human self-awareness. What they do extraordinarily well is process data and use prediction to come out with results. They do not think, they process data. There is a distinction and this is what sets machines apart from humans. We think, they don't. That sounds simple, but the scale is extraordinary. If someone starts the sentence, "The capital of Nigeria is…," you already know the next word is likely to be "Abuja" because your brain has learned patterns. Large Language Models operate in a similar way, but on a massive scale. They predict sequences of words, one after another, thousands of times per second. The result is language that often feels remarkably human.

What Does GPT Mean?

Since ChatGPT is one of the most widely known AI tools, it is worth understanding its name. GPT stands for Generative Pre-trained Transformer.

- Generative means it can create content.
- Pre-trained means it has already been trained on vast amounts of information before you begin using it.
- Transformer refers to the architecture that make modern language models possible.

The transformer architecture was introduced in the 2017 paper *Attention Is All You Need* by Ashish Vaswani, Noam Shazeer, Niki Parmar, Jakob Uszkoreit, Llion Jones, Aidan N. Gomez, Łukasz Kaiser, and Illia Polosukhin. The authors wrote, "We propose a new simple network architecture, the Transformer, based solely on attention mechanisms, dispensing with recurrence and convolutions entirely." This architecture became a foundation for modern language models because it made sequence processing more efficient. (Vaswani, 2017).

The Breakthrough That Changed Everything

For many years, computers struggled with language. They could process words, but understanding context was very difficult. Transformer models changed that by helping AI systems understand relationships between words, sentences, and concepts more effectively. This breakthrough accelerated the development of major modern AI systems, such as, ChatGPT, Claude, Gemini, Grok.

What Are Tokens?

Another term you will hear frequently is **tokens**. A token is a chunk of text that an AI model processes. It may be a full word, part of a word, a single character, a space, or punctuation. For example, OpenAI explains that models process text in tokens rather than full human sentences, and that they represent sequences of characters and are the building blocks AI uses to read and generate language. OpenAI gives a useful rule of thumb. One token is roughly four characters of English text, or about three-quarters of a word. That means 100 tokens is approximately 75 English words, although the exact number varies by language, wording, punctuation, and model. AI systems have limits on how many tokens they can process at once, and many AI services also measure usage by tokens. (OpenAI, What are tokens and how to count them?, n.d.)

What Is Generative AI?

Generative AI refers to systems that create new content rather than simply retrieving existing information. That content can include:

- Text, reports, presentations, and business plans
- Images, videos, music, and audio
- Code, marketing and creative concepts

This is a major shift. Traditional software typically retrieves information. Generative AI creates new content based on patterns it has learned, and that capability is already transforming industries.

The Good News for Executives

When executives first hear about AI, they often worry that they need to become technical experts. They do not, but they do need to become informed leaders. Most CEOs do not build computer processors, but they understand technology strategy, much like hospital administrators are not surgeons, but they understand healthcare operations, or airline executives that do not fly aircraft, but they understand aviation. Similarly, leaders do not need to build AI systems, but they do need to understand what those systems can and cannot do. That understanding is becoming a critical leadership skill.

When AI Hallucinates

At this point, we need to discuss one of the most important limitations of Large Language Models which is that sometimes they are wrong. Very wrong. And they can be wrong with impressive confidence. In the AI world, this is often called a hallucination. A hallucination occurs when an AI system generates information that sounds plausible but is factually incorrect. The AI is not lying or trying to deceive, rather it is predicting language patterns, and sometimes those predictions produce inaccurate information.

This is why verification remains essential. Human Oversight is a necessity. Never assume that AI-generated content is automatically correct. Always verify critical information, especially in high-stakes environments such as law, healthcare, finance, engineering, and government. In Prompting which we will go into later, you can ask the AI system to only produce or provide factual information. Ask not to make up information, or add details that are not based on facts.

OpenAI's *GPT-4 Technical Report* explains that GPT-4 is "a Transformer-based model pre-trained to predict the next token in a document" and notes that post-training improved performance on measures of factuality. The report also makes clear that model outputs still require caution because improved factuality does not mean perfect accuracy. (OpenAI, GPT-4 Technical Report, .2023).

The National Institute of Standards and Technology identifies generative AI as presenting unique risks and released the *Artificial Intelligence Risk Management Framework: Generative Artificial Intelligence Profile* to help organizations identify and manage those risks. (technology, 2024).

The Human-AI Partnership

One of the biggest misconceptions about Generative AI is that it will replace human beings entirely. However, calculators have not eliminate accountants. Spreadsheets have not eliminate finance professionals. Email did not eliminate communication. The internet did not eliminate research. Instead, these tools changed how people work. Generative AI is likely to do the same. The winners will not necessarily be the people with the most technical knowledge, but those who understand how to combine human judgment with artificial intelligence.

My View from the Boardroom

As I delivered AI awareness sessions to executives, professionals, and organizations, I noticed a common pattern. The most successful leaders are not asking, "Will AI replace us?" They are asking, "How can we use AI responsibly and effectively?" That is the right question, because AI is not going away. It is becoming more capable, more accessible, and more integrated into daily life. The real challenge is not deciding whether AI matters. The real challenge is learning how to use it wisely. Understanding AI is one thing. Leading with AI is something else entirely. And that is where we are headed next.

Chapter 10

AI in the Boardroom - What Leaders Need to Know

One of the strengths of this book is that it does not approach Artificial Intelligence only from theory. It is written from lived experience, studying AI at Oxford, attending lots of international AI summits featuring leading AI experts, completing hundreds of hours of research, speaking with people outside the classroom, and helping executives, professionals, and organizations understand what AI means for their work.

The Oxford Assignment That Changed My Perspective

One of the most memorable assignments during my Oxford studies was when our professors sent us into the city with a simple task and that was to talk to people about what they knew about Artificial Intelligence, how they felt about it, and whether they believed it would affect their lives or careers.

The exercise seemed straightforward, but it became one of the most revealing experiences of the entire program. As my classmates and I spoke with people from different professions and backgrounds, a clear pattern emerged. People were thinking about AI, but they were thinking about it in very different ways. Some people were excited. Some were skeptical. Some were afraid of the potential for great harm. But none was completely unconcerned. Those reactions taught me an important lesson, which is that AI is not only a technology conversation, but it is also a human conversation.

When Disruption Feels Far Away

One woman we spoke with worked as a receptionist. When we asked whether she believed AI might affect her job, she confidently said no. At the time, her answer reflected what many people believed and it was that AI was something that would happen somewhere else, to someone else, in another industry, not in their own workplace or daily life. Looking back, we can see how quickly the landscape changed. Very quickly. Many functions traditionally associated with reception, scheduling, customer inquiries, appointment management, information retrieval, and administrative support are now being assisted by AI systems. The lesson was not that she was wrong. The lesson was that most of us struggle to see disruption before it arrives.

Fear, Hype, and the Need for Understanding

Another group expressed a very different concern. They were convinced AI would eventually take over the world. Some worried about robots. Others worried about machines becoming smarter than humans. Several fears sounded more like science fiction than business strategy. What struck me was that very few people understood how AI actually worked. Many opinions had been shaped by headlines, movies, and media coverage. Fear often grows in the absence of understanding. The more people understand AI, the less likely they are to view it through extremes. Neither panic nor blind optimism is helpful. Understanding is.

The Creative Work Surprise

Perhaps the most memorable lesson came from inside our classroom. During one discussion, a senior executive was asked whether Artificial Intelligence would significantly affect graphic design and creative departments. Her answer was an immediate "no." Her basis was that creativity was considered a human domain, and the prevailing assumption was that AI might automate repetitive tasks but would not produce creative work. This reality has changed.

Today, AI can generate logos, illustrations, marketing concepts, advertising copy, edited photographs, videos, presentations, music, and digital artwork. It has not eliminated graphic designers, but it has changed the profession. Designers who learn to use AI can often produce work faster and more efficiently, while those who ignore it may find themselves competing against professionals who can do more in less time.

The Boardroom Lesson

Throughout history, technology has rarely eliminated entire professions overnight. More often, it changes how work is performed. Professionals who adapt tend to thrive and those who refuse to adapt often struggle. That is one of the most important lessons executives need to understand. The question is no longer whether AI will affect your industry. It will. The better questions are:

1) Where will AI create the greatest opportunity for my organization?

2) Where will AI create the greatest risk?
3) What must leaders understand now so they can guide the organization responsibly?

Those are leadership questions. They are boardroom questions. They are strategy questions. They require leaders who understand enough about Artificial Intelligence to ask better questions, evaluate trade-offs, and make informed decisions.

AI Literacy Is Becoming a Leadership Competency

One of the other reasons I wrote this book is because I observed a troubling gap. Many executives are expected to make decisions about AI without fully understanding its capabilities, limitations, risks, or opportunities. That is like asking someone to drive a vehicle without understanding the dashboard or how to apply the breaks or navigational system. AI literacy is rapidly becoming a leadership competency, not because every executive needs to become an engineer or every board member needs to learn programming, but because every leader needs to understand enough about AI to ask intelligent questions and make better decisions. The leaders who thrive will not be the ones who know everything about AI. They will be the ones who learn quickly enough to lead responsibly while the technology continues to evolve.

Chapter 11

Will AI Take My Job?

If I had to choose the question people ask me more than any other about Artificial Intelligence, it would be," Will AI take my job?" I have heard this question from executives, lawyers, doctors, teachers, government officials, students, business owners, and people who had never previously paid attention to technology. The question is universal because work is not only how we earn money. It is tied to identity, purpose, dignity, and the ability to provide for our families.

A Question From the Legal Profession

I remember one of the first AI awareness sessions I conducted with my fellow cohort member, Johannes delivered to WIMBIZ, Women in Management, Business and Public

Service. During the session, an attorney asked the question perhaps a few were thinking, "Is AI going to take my job?" At the time, stories about AI and the legal profession were generating headlines. If AI could review documents, conduct research, and assist with legal processes, people naturally wondered what would happen to attorneys.

My answer then has changed from the answer I would give today. AI is unlikely to replace lawyers entirely, but it is already changing how lawyers work. Today, attorneys use AI for legal research, contract review, document summarization, case preparation, and other tasks. Yet clients still need legal judgment, negotiation, strategy, and advocacy. Technology can assist those functions, but it does not automatically replace them, at least not for now.

A Prediction I Got Wrong

I once believed certain creative professions would remain largely untouched by Artificial Intelligence. Painting was one example. Creative expression felt deeply human. Art felt safe. Many of us assumed a machine could not replicate imagination. Today, AI systems can create paintings, digital artwork, illustrations, concept art, graphic designs, marketing visuals, and more, in seconds. Instead of asking which jobs are completely safe, a better question is, "how will AI change the work we do?" Change is far more common than complete replacement.

What AI Does Best and What Humans Still Do Best

Artificial Intelligence tends to perform well in areas involving:

1) Pattern recognition and prediction
2) Data analysis and information retrieval
3) Repetitive activities and large-scale processing
4) Routine decision support

Humans continue to excel in areas involving:

1) Empathy, trust, and relationship building
2) Ethics, judgment, and common sense
3) Leadership, negotiation, and strategic thinking
4) Complex decision-making

As AI becomes more capable, uniquely human abilities become more important, not less. The future will not belong to people who compete with AI on speed. It will belong to people who combine human judgment with AI-enabled capability.

The Real Threat

The greatest threat is not AI itself. The greatest threat is standing still while the world changes around us. The professionals most likely to struggle are not necessarily those in a particular industry, but those who refuse to learn, adapt, or those who believe that tomorrow will be the same as yesterday.

The Better Question

Over time, the question of "Will AI take my job?" will evolve into: "How can I use AI to become better at my job?" That question shifts the conversation from fear to opportunity, from victimhood to leadership, and from uncertainty to preparation. Humans who learn to work effectively with AI will often outperform humans who do not. That may be the most important workforce lesson of the AI era.

The Legal Profession and the Value Question

As my AI awareness work continued, I had the opportunity to present AI to members of the Nigerian Bar Association on more than one occasion. Not surprisingly, attorneys asked many of the same questions I heard from other professionals. Would AI reduce opportunities for young attorneys? Would it make legal research obsolete? Would it eventually argue cases better than humans? One question stood out because it went beyond technology and into business models. If attorneys traditionally bill by the hour, and AI allows a task that once took an hour to be completed in a few minutes, can attorneys still charge the same fee?

What is the ethical answer? It is an excellent question because it forces us to rethink how value is defined. For generations, many professional services like law, consulting, accounting, and others have been priced according to time. The more hours worked, the more revenue generated. AI challenges that model.

Imagine a lawyer who previously spent five hours researching an issue. Today, AI may help gather relevant information in a fraction of the time. The attorney still applies professional judgment, validates the information, advises the client, and carries responsibility for the outcome. But the research process may be dramatically faster. That raises difficult but important questions:

1) Is the client paying for the time spent or the expertise applied?
2) Is the value in the effort, the outcome, or the risk reduced?
3) How should professionals price work when AI accelerates part of the process?

AI is forcing many professions to rethink the relationship between time and value. The reality is that clients do not pay only for hours. They pay for knowledge, experience, judgment, accuracy, insight, risk reduction, and results.

While AI may reduce the time required to complete certain tasks, it does not automatically eliminate the value of expertise. In fact, as routine work becomes more automated, professional judgment may become even more valuable. This question will extend far beyond law. Consultants, accountants, marketers, financial advisors, educators, healthcare professionals, and many other knowledge workers will need to reconsider how value is created, measured, and compensated in an AI-enabled world.

From Understanding to Implementation

Up to this point, the reader has learned what AI is, how it evolved, why it matters, how it may affect jobs, and how executives should think about it. Now we need to answer the next practical question every executive, entrepreneur, attorney, doctor, educator, nonprofit leader, and business owner is asking: "Okay, I understand AI. Now what do I actually do with it?" This is where this book moves from awareness to actionable. Many books stop at theory. The next section moves into implementation.

Chapter 12

Understanding AI & Using AI - A Practical Guide for Professionals and Executives

If you have made it this far in the book, congratulations. You now understand what Artificial Intelligence is, where it came from, how it works, and why it matters. You also understand that AI brings both opportunity and risk. The next question is the practical one. "What should I actually do with it?" This is where some professionals get stuck. They attend conferences, watch videos, read articles, and listen to experts, then return to work and do nothing differently. The truth is that knowledge without action creates very little value. Artificial Intelligence becomes useful only when it is applied to real problems.

Three Common Responses to AI

Over the years, I have noticed that people generally respond to AI in three ways:

1) They assume AI is a passing trend and hope it will disappear.
2) They become overwhelmed. They see too many tools, announcements, and developments, and they do not know where to begin, because it seems like a lot, and it is a lot, but AI can also simplify the learning process.
3) They experiment. These are the people who learn the fastest, not because they know the most, but because they are willing to try. Like me.

I encourage you to become part of the third group. You do not need to become an expert overnight. You simply need to start.

Start With the Problem, Not the Technology

My first rule of AI adoption is simple. Do not start with the technology. Start with the solution you would like to see for a problem. One of the biggest mistakes organizations make is becoming excited about AI and immediately searching for tools. That is backwards. Before choosing a platform, ask practical questions:

1) What problem am I trying to solve?
2) What task takes too much time?
3) What process is repetitive?
4) What information takes too long to analyze?
5) Where do bottlenecks slow down the work?

A Five-Minute AI Exercise

Whenever I work with professionals who are new to AI, I ask them to complete a simple exercise. Draw three columns and list:

1) Tasks I perform repeatedly
2) Tasks that consume significant time
3) Tasks I dislike doing

You will often discover that many items on those lists can be partially automated or assisted by AI, but the professional remains responsible for the final output. AI assists while the human decides. That partnership is where the greatest value exists.

Start Small and Scale Thoughtfully

One of the biggest misconceptions about AI implementation is that it requires a massive budget. Sometimes it does, but often it does not. Many professionals and organizations can begin with tools they already have access to. The key is to start with one practical use case, apply it consistently, learn from the results, improve the process, and then expand. Organizations that succeed with AI usually start small and scale thoughtfully. Organizations that fail often try to transform everything at once. A focused beginning creates learning, confidence, and momentum.

Learn to Prompt Clearly

If Artificial Intelligence is the engine, prompting is the steering wheel. A prompt is simply an instruction, but the quality of that instruction often determines the quality of the result. Many people type one vague sentence into an AI tool, receive a generic answer, and conclude that the technology is unimpressive. Often, the problem is not the AI, but rather, it is the instruction. The clearer your request, context, audience, format, and desired outcome, the better your results are likely to be. Prompting is quickly becoming one of the most valuable workplace skills of the AI era because it teaches people how to communicate with intelligent systems in a way that produces useful work.

Build a Personal AI Toolkit

One of the most practical steps you can take is to create a personal AI toolkit. The goal is not to collect dozens of tools. The goal is to identify a small number of tools that solve real problems in your work such as:

1) Writing, research, and presentations
2) Data analysis, meetings, and project management

3) Image creation, video creation, learning, and communication

The Most Important Lesson

After teaching AI awareness sessions and speaking with executives, attorneys, nonprofit leaders, educators, and professionals, I have reached a simple conclusion. The people who benefit most from AI are not necessarily the most technical, but the most curious who experiment, learn, adapt, ask questions, and remain open to change. Artificial Intelligence is not a destination. It is a journey. The best time to begin that journey is now, because while some people are still debating whether AI matters, others are already using it to become more productive, creative, and effective and the gap between those two groups is growing every day.

The Moment I Realized Awareness Was Not Enough

A few years ago, I conducted an AI awareness session for the Center for Public Service Productivity and Development, (CeProd). Like many of my early sessions, the goal was to help participants understand what AI is, why it matters, and how it might affect careers, organizations, and society. The session was well received. Participants asked thoughtful questions, the discussion was lively, and then the organizer approached me and said something that changed how I thought about AI education. "This was helpful. But now we need another class. We need a class that teaches us how to actually implement AI."

He was absolutely right. Understanding AI and implementing AI are two different things. Awareness, education, and history are important, but eventually people stop asking, "What is AI?" and begin asking, "How do I use it?" That is the more powerful question.

Organizations do not create value simply because employees understand Artificial Intelligence. They create value when employees apply Artificial Intelligence to solve real problems. That conversation helped me see that many organizations move through five stages of AI maturity:

1. Awareness: Learn about AI and become curious.
2. Understanding: Begin learning the concepts.
3. Experimentation: Test AI tools and explore use cases.
4. Implementation: Integrate AI into real workflows.
5. Transformation: Redesign how work gets done.

Many organizations never move beyond awareness. They attend seminars, listen to speakers, read articles, and discuss AI, but nothing changes. That participant's comment became one reason I expanded my own approach to AI education. People no longer wanted awareness alone. They wanted application, implementation, and results. In the age of Artificial Intelligence, knowing is no longer enough. The real advantage belongs to those who apply what they know.

Chapter 13

Data: The Fuel That Powers Artificial Intelligence

Artificial Intelligence starts with data. Before an AI system can recognize patterns, make predictions, generate insights, or support decision-making, it needs something to learn from. Data.

Throughout my AI awareness sessions, I often ask audiences, "What do you think is the most valuable asset in your organization?" Some say people. Others say technology. All of those are important, but in today's digital economy, data has become one of the most valuable organizational assets.

The Day I Walked Into a Law Library

During one AI engagement where I had gone to deliver a talk on AI at the Law Faculty, I found myself looking at shelves filled with legal materials, books, case reports, research papers, legal opinions, reference materials, and decades of accumulated knowledge. The collection was impressive. It represented enormous intellectual capital. But it also raised an important question. How much had been digitized, made searchable, organized for analysis, or prepared to support future legal research and AI?

Many organizations possess enormous amounts of information, but much of it remains trapped in filing cabinets, storage rooms, paper documents, boxes, physical libraries, or disconnected systems. Before organizations can benefit fully from AI, they must first make their information accessible.

Data Versus Information

People often use the terms data and information interchangeably, but they are related rather than identical. Data is the raw material. Information is data that has been organized and given meaning, for example, a spreadsheet containing thousands of customer transactions is data. A report showing which products generate the most revenue is information. A recommendation about what to promote next quarter is insight. Artificial Intelligence helps organizations move from data to information, to insight more quickly than ever before.

Why AI Needs Data

Imagine trying to teach a child mathematics without numbers or reading without words. It would be impossible. The same principle applies to Artificial Intelligence. AI learns from examples, and those examples come from data.

Those examples may come from customer, financial, medical, legal, educational, manufacturing, or agricultural data. The more relevant and reliable the data, the better the AI system can perform. This is why data quality matters so much. Poor data often produces poor results.

The Digital Transformation Challenge

For many organizations, the first step toward AI is not purchasing software, it is digitization. That means converting paper records into digital records, organizing files, creating searchable databases, improving data quality, establishing standards, and building governance processes. This work may not sound exciting, but it is essential. Many AI projects fail not because the AI is weak, but because the underlying data is incomplete, inaccurate, inconsistent, or inaccessible.

Where Does the Data Live?

Eventually, organizations must answer another practical question. Where will the data live? Some organizations use cloud services, others use private data centers, and many use a hybrid approach. Regardless of the storage model, leaders must consider reliability, cybersecurity, backup systems, disaster recovery, physical security, data privacy, and continuity of operations. These may sound like technology issues, but they are business issues. If data is unavailable, compromised, or lost, operations, reputation, and institutional memory may suffer.

The Role of the Chief Data Officer

As organizations become more data-driven, many are creating leadership roles dedicated to data management. One such role is the Chief Data Officer, or CDO. The CDO helps the organization manage, govern, protect, and derive value from data. Although the role varies by organization, common responsibilities include:

1) Data governance, quality, privacy, and compliance
2) Data strategy, architecture, analytics, and AI readiness
3) Connecting technology teams with business leaders

What Qualifications Does a Chief Data Officer Need?

There is no single path to becoming a Chief Data Officer. Some leaders come from information technology, analytics, business strategy, or data science. The strongest CDOs often combine technical knowledge with business understanding because data is not merely a technology issue; it is a strategic business asset.

Did AI Replace the Chief Data Officer?

When people hear about AI automating reports, dashboards, and routine analysis, they sometimes wonder whether it will replace data leadership roles. The answer is more complicated than a simple yes or no. In most organizations today, AI is not replacing Chief Data Officers. If anything, it is increasing the need for effective data leadership. What is changing is the nature of the role. Routine reporting, basic analytics, data visualization, and some administrative functions can increasingly be automated. But organizations still need leaders who can answer strategic questions about what data to collect, how to govern it, how to protect it, how to use it ethically, and how to align data strategy with business strategy.

The Real Lesson

When organizations talk about Artificial Intelligence, they often focus on the technology. The smarter conversation begins earlier. It begins with data. Before asking, "What AI tool should we buy?" leaders should ask, "What data do we have?" and instead of, "What model should we use?" they should ask, "Can we trust our data?" not "How do we implement AI?" but they should ask, "Are we ready?" Every successful AI journey begins with data.

Is Data the New Oil?

During my studies at Oxford and in many AI discussions since then, I repeatedly heard the phrase, "Data is the new oil." At first glance, the comparison makes sense. Oil powered the Industrial Revolution while data is helping power the Artificial Intelligence revolution. Oil created economic value, data is creating economic value. Oil fueled industries, data fuels AI.

But the comparison is not perfect. Crude oil has little value until it is refined, and data is similar. Raw data sitting in a filing cabinet, spreadsheet, hard drive, or database does not automatically create value. It must be organized, cleaned, structured, analyzed, protected, governed, and transformed into useful insight. In that sense, data may be even more powerful than oil. Oil becomes fuel. Data can become knowledge, intelligence, prediction, decision-making, and innovation, powering current and future digital economies.

As I travel to AI conferences around the world, I often notice the same pattern. I receive a complimentary pen, mug, notebook, and then I am asked to provide my name, email address, company, job title, or other personal information. At first glance, it appears to be a simple exchange. I receive a small gift, and the organization collects my contact details. However, the real value is rarely the pen or the mug. The real value is the information provided.

The same thing happens online. Many websites and AI platforms ask whether users are willing to share their information to improve products, personalize their experience, measure marketing performance, or deliver more relevant content. While these requests are often presented as settings or preferences, they also demonstrate an important business principle and this is that data has value.

For business leaders, this should not come as a surprise. Organizations collect data because it helps them better understand their customers, identify trends, improve products and services, measure the effectiveness of marketing campaigns, and make better business decisions. Artificial Intelligence becomes more useful when it is developed and deployed using high-quality, relevant, and responsibly managed data.

This is one of the reasons data is often compared to oil. Crude oil has little value until it is refined into products such as fuel, plastics, and chemicals. In much the same way, raw data becomes valuable only after it has been collected, organized, cleaned, analyzed, and transformed into meaningful insights. There is, however, an important difference. Oil becomes depleted as it is consumed. Data does not. The same dataset can be analyzed repeatedly, combined with other datasets, and used to solve entirely different business problems. When managed responsibly and ethically, data becomes a renewable strategic asset whose value often grows over time.

As CEOs, we should ask an important question in line with "what value does this data create, and how can my organization use its own data responsibly to create better products, better services, and better decisions?" That is where the real competitive advantage lies.

In AI platforms, if you do not want the data you input to be used as training data for Large language Models, please go to the settings page and then click on Data Controls. Here Users may choose whether the information they share can be used to help improve Artificial Intelligence systems. When permission is granted, conversations and interactions may be analyzed to enhance the quality, accuracy, and usefulness of future AI models. Organizations that collect this information are expected to implement safeguards designed to protect users' privacy and sensitive information.

Measuring Marketing Performance

Some organizations use cookies and other analytical tools to understand how effective their marketing activities are. This information helps them evaluate which campaigns are reaching the right audiences and where improvements can be made.

Personalized Marketing

Businesses may also use data to provide more relevant marketing based on a user's interests, preferences, or previous interactions. The goal is to deliver communications and recommendations that are more meaningful to each individual rather than presenting the same message to everyone.

Advertising Preferences

Many digital platforms allow users to decide whether advertisements should be tailored to their interests. Giving people control over these settings helps them choose how their information is used while promoting greater transparency.

Chapter 14

Understanding Data Analytics: The Foundation of Intelligent Decision-Making

At its most fundamental level, data analytics is the process of examining data to identify patterns, trends, relationships, and insights that support informed decision-making. While the concept may initially appear technical, its practical application is both straightforward and highly relevant to everyday business operations.

Consider the example of a supermarket. Every day, customers purchase products, generating thousands of transactions. Each transaction creates valuable information about consumer behavior, purchasing patterns, inventory and operational performance. On its own, this information has limited value. However, when analyzed effectively, it can provide answers to critical business questions. A retailer may seek to understand which products sell most frequently, which items generate the least demand, and what times of day experience the highest customer traffic.

Management may wish to identify the customers who spend the most money, determine which products are commonly purchased together, or decide whether certain items should be stocked in greater quantities while others should be discontinued. Data analytics transforms raw information into actionable insights. These insights support better decisions, and better decisions ultimately drive improved organizational outcomes.

The Four Levels of Analytics

Throughout my studies of Artificial Intelligence and data-driven decision-making, as well as discussions with leaders across various industries, I have found that most analytical activities can be categorized into four distinct levels. Each level answers a different business question and contributes to a deeper understanding of organizational performance.

Descriptive Analytics: What Happened?

Descriptive analytics represents the most basic and widely used form of analytics. Its purpose is to summarize historical information and provide a clear picture of past events. Common examples include sales reports, financial statements, monthly performance summaries, and executive dashboards. These tools present information such as revenue generated during a specific period, the number of customers served, the volume of legal cases completed, the number of patients treated, or student enrollment figures.

Descriptive analytics tells the story of the past and it also answers the question, "what happened?" by presenting factual information about completed activities and outcomes.

Diagnostic Analytics: Why Did It Happen?

While descriptive analytics explains what occurred, diagnostic analytics seeks to understand why it occurred. Suppose a company discovers that sales have declined by twenty percent. Descriptive analytics identifies the decline, but diagnostic analytics investigates the underlying causes. The decline may have resulted from increased competition, changes in pricing, shifts in customer preferences, supply chain disruptions, or broader economic conditions. By identifying the factors that contributed to a particular outcome, diagnostic analytics moves organizations beyond simple observation toward genuine understanding. It enables leaders to address root causes rather than merely react to symptoms.

Predictive Analytics: What Is Likely to Happen Next?

Predictive analytics uses historical data, statistical techniques, and machine learning models to estimate future outcomes. Although predictions are never perfect, they can significantly improve decision-making by reducing uncertainty.

Organizations across many sectors rely on predictive analytics. Financial institutions use it to estimate the likelihood of loan default. Healthcare providers use predictive models to identify patients who may be at greater risk of complications. Insurance companies estimate claims risk, universities identify students who may require additional support, and retailers forecast future demand for products and services.

The objective of predictive analytics is not absolute certainty. Rather, it is to provide decision-makers with a more informed view of possible future events. Even imperfect forecasts can offer valuable guidance when compared with making decisions without any predictive insight.

Prescriptive Analytics: What Should We Do?

Prescriptive analytics represents the most advanced stage of the analytical process. It not only predicts potential outcomes but also recommends actions that may produce the most desirable results. For example, a logistics company may use prescriptive analytics to determine the most efficient delivery routes. A hospital may optimize the allocation of medical staff and resources. Manufacturers may improve production schedules, airlines may refine ticket pricing strategies, and investment firms may recommend portfolio adjustments based on changing market conditions.

Prescriptive analytics transforms insight into action. It bridges the gap between understanding a situation and determining the most effective response. This is often where the greatest business value emerges.

A Faculty of Law Example

An illustration of the power of analytics can be found within academic institutions, particularly law faculties and legal research centers. Many law libraries contain decades of accumulated knowledge, including case law, legal opinions, research publications, legislative materials, and academic scholarship. Traditionally, much of this information has been stored in physical collections or fragmented digital systems, making access and analysis time-consuming.

Imagine if these resources were fully digitized, organized, categorized, tagged, and made searchable. Analytics could then be applied across decades of legal information. Researchers could identify patterns in judicial decisions, analyze legal trends over time, locate relevant precedents more efficiently, and accelerate complex legal research. The value does not lie merely in possessing information. The true value lies in making that information accessible, searchable, as well as usable. This principle extends far beyond the legal profession and applies to virtually every industry.

Why Many Organizations Struggle with Artificial Intelligence

One of the most common reasons organizations struggle with Artificial Intelligence initiatives is that they attempt to implement AI before establishing a strong analytical foundation. This approach is comparable to constructing the second floor of a building before completing the foundation. Artificial Intelligence systems depend on high-quality data and meaningful analytical processes. They cannot compensate for poor data quality, missing information, or a lack of understanding about the underlying data. Before investing heavily in AI technologies, organizations should ask several fundamental questions:

1) Do we understand our data?
2) Do we trust the accuracy and quality of our data?
3) Are we analyzing our data effectively?
4) Do we know what business questions we are trying to answer?

The answers to these questions are often more important than the selection of any specific AI platform or software solution.

From Reporting to Prediction

Many organizations devote substantial resources to reporting historical information. Far fewer invest the same effort in predicting future outcomes, and even fewer focus on recommending actions based on those predictions.

Yet this progression, from reporting to understanding, understanding to prediction, and prediction to action, represents the future of data-driven organizations.

The organizations that will thrive in the era of Artificial Intelligence will be the organizations that consistently make better decisions, more quickly and with greater confidence. Better decisions are driven by better insights. Better insights come from effective analytics. Effective analytics depends on high-quality data. The journey invariably begins with data.

Executive Takeaway

When senior executives ask where they should begin their Artificial Intelligence journey, the first step is not immediately purchasing AI software, hiring teams of data scientists, or launching large-scale AI initiatives. Instead, the starting point is first to determine what problems need solving then having a thorough understanding of the organization's data. Once data is understood and trusted, meaningful analytics becomes possible. Once analytics is established, the result is that prediction becomes achievable. Once prediction becomes achievable, more intelligent and proactive decision-making can follow. This progression, from data to analytics, from analytics to prediction, and from prediction to intelligent action, is where Artificial Intelligence begins to create genuine and sustainable value for organizations.

Chapter 15

Building an AI-Ready Organization

Moving From Curiosity to Execution

One of the most common mistakes organizations make is believing that implementing Artificial Intelligence must always start with new technology. This is often the last piece of the puzzle, not the first. By the time an organization starts evaluating AI tools, selecting vendors, or testing software, a great deal of preparation should already have taken place in thinking strategically about how AI fits into their mission, operations, workforce, and long-term objectives. So CEOs, first hold off on the purchase orders. First ask the questions in the previous chapters.

Throughout my AI awareness sessions, I have observed a recurring pattern. Leaders become excited about Artificial Intelligence after attending a conference, reading an article, or hearing about a competitor's success. They immediately begin asking questions about which AI platform to buy or which chatbot to implement. While those questions are understandable, they are often premature.

The first question should not be, "What AI tool should we purchase?"

The first question should be, "What business problem are we trying to solve?"

Organizations that begin with technology often struggle. Organizations that begin with clearly defined business objectives are far more likely to succeed. For example, if customer service response times are creating frustration for clients, an AI-powered chatbot may be an appropriate solution, or when employees are spending excessive time searching for information buried inside company documents, an internal AI knowledge assistant or Agents may be useful. In repetitive reporting tasks, automation tools create significant efficiencies. The goal is to solve meaningful business problems.

One of the most visible forms of AI implementation today is the chatbot and agentic AI. Most people are already familiar with customer-facing chatbots that appear on websites and answer basic questions. These systems have become dramatically more capable over the last few years. Rather than relying on a fixed set of scripted responses, modern AI-powered chatbots and Agents can understand context, respond conversationally, and assist customers with increasingly complex inquiries.

Imagine a new employee joining an organization. Instead of searching through hundreds of policy documents, training manuals, meeting notes, and procedures, the employee could simply ask an internal AI assistant a question and receive an immediate answer.

"What is our travel policy?"
"Where can I find the expense reimbursement form?"
"What are the requirements for vendor approval?"
"What projects are currently underway?"

Today, AI-assisted development tools are enabling organizations in building applications faster than ever before. The barriers to innovation are falling rapidly. Organizations that once lacked the resources to build custom solutions are finding that AI can help accelerate development and reduce costs. Perhaps the most exciting development on the horizon is Agentic AI. If chatbots answer questions, Agentic AI acts. This distinction is important. Traditional AI systems respond when prompted.

Agentic AI systems can perform multi-step tasks with a degree of autonomy while still operating within boundaries established by humans. Imagine instructing an AI agent to organize a conference. Rather than simply providing recommendations, the system might identify venues, compare pricing, draft communications, coordinate scheduling, track responses, and generate status reports. A human remains responsible for oversight and final decisions, but the AI assists with execution. Similarly, an AI agent supporting a legal practice might monitor regulatory changes, summarize updates, draft preliminary reports, and alert attorneys when significant developments occur.

An AI agent supporting a nonprofit organization might monitor grant opportunities, prepare application drafts, track deadlines, and provide progress updates. These capabilities are moving us beyond simple automation and into a world where AI increasingly functions as a digital collaborator. That said, leaders should approach Agentic AI thoughtfully. Just because a task can be automated does not mean it should be fully delegated. Human judgment remains essential, particularly in areas involving ethics, compliance, governance, legal responsibility, and strategic decision-making.

A Step-by-Step Guide to Creating an Agentic AI

An agentic artificial intelligence (AI) system is designed to pursue goals by making decisions, planning actions, using external tools, and adapting its behavior based on new information. Unlike a traditional chatbot that primarily responds to prompts, an agentic AI can execute a sequence of tasks in order to accomplish an objective while operating within predefined rules and constraints.

Creating an agentic AI involves integrating several technologies, including a large language model (LLM), memory, external tools, planning mechanisms, and safety controls. The following sections describe this process in a logical sequence.

Step 1: Clearly Define the Purpose of the Agent

The first step is to determine exactly what the AI agent is intended to accomplish. Every design decision that follows depends on a clearly defined purpose.

Questions to answer include:

- What problem will the agent solve?
- Who will use the agent?
- What tasks should the agent perform independently?
- What tasks should require human approval?
- What limitations should the agent have?

For example, an AI research assistant may gather information, summarize documents, and organize notes, while a customer service agent may answer questions, retrieve account information, and escalate complex issues to a human representative.

A well-defined objective helps prevent unnecessary complexity and establishes measurable criteria for success.

Step 2: Select an Appropriate Large Language Model

The language model serves as the reasoning component of the agent. It interprets instructions, analyzes information, generates responses, and determines which actions should be taken.

When selecting a model, consider factors such as:

- Reasoning ability
- Context window size
- Cost
- Response speed
- Availability through an application programming interface (API)
- Support for structured tool calling

Different models have different strengths. The appropriate choice depends on the specific requirements of the project.

Step 3: Design the Agent's Instructions

Every agent requires clear instructions.

These instructions define:

- The agent's role
- Its responsibilities
- Behavioral expectations
- Safety requirements
- Communication style
- Decision-making boundaries

For example, an agent designed for financial analysis should be instructed to identify uncertainty, avoid unsupported conclusions, and recommend human review before important financial decisions are made.

Clear instructions improve consistency and reduce unintended behavior.

Step 4: Design the Agent's Memory

Memory enables an agent to maintain context over time. Memory is commonly divided into three categories.

Short-Term Memory

Short-term memory stores information relevant to the current interaction and allows the agent to remember previous messages and maintain a coherent conversation.

Long-Term Memory

Long-term memory stores information across multiple sessions. This may include user preferences, recurring projects, or previously completed tasks.

Knowledge Memory

Knowledge memory consists of documents, manuals, databases, or other reference materials that the agent can retrieve when answering questions or making decisions.

Different memory systems serve different purposes, and many production systems combine all three.

Step 5: Provide Access to External Tools

Language models generate text, but they cannot directly interact with external systems unless tools are provided.

Examples of tools include:

- Internet search
- Database queries
- Calendars
- Email services
- File systems
- Programming environments
- Business software
- Application programming interfaces (APIs)

The agent determines when a tool is needed, submits the appropriate request, receives the result, and incorporates that information into its reasoning process.

Tool integration significantly expands the capabilities of an AI agent beyond text generation.

Step 6: Implement a Planning Process

Agentic systems generally accomplish complex goals by dividing them into smaller tasks. Rather than attempting to solve a problem in a single response, the agent follows a structured planning process.

A typical planning cycle includes:

1. Understanding the objective.
2. Identifying the required tasks.
3. Determining the order in which those tasks should be completed.
4. Executing each task.
5. Evaluating the outcome.
6. Revising the plan if necessary.

This iterative process enables the agent to adapt when unexpected situations occur.

Step 7: Enable Decision-Making

An agentic AI must evaluate available information before choosing its next action. Decision-making often involves questions such as:

- Is additional information required?
- Which tool should be used?
- Has the objective already been achieved?
- Should user be asked for clarification?
- Is human approval required before continuing?

Well-designed decision logic improves both accuracy and efficiency.

Step 8: Incorporate Retrieval-Augmented Generation (RAG)

Many agentic systems use Retrieval-Augmented Generation (RAG) to improve factual accuracy. Rather than relying solely on information contained within the language model, the system first searches relevant documents or databases. The retrieved information is then supplied to the language model as additional context before a response is generated.

This approach helps the agent provide answers that are grounded in current or organization-specific information.

Step 9: Add Feedback and Self-Evaluation

Effective agents evaluate the results of their actions. After completing a task, the agent should determine whether the objective has been achieved.

If the outcome is unsatisfactory, the agent may:

- Try an alternative strategy.
- Use a different tool.
- Request additional information.
- Repeat process with revised parameters.

Feedback loops are inherently designed to improve reliability, particularly during multi-step workflows.

Step 10: Establish Safety Constraints

Every autonomous system should operate within clearly defined limits.

Examples include:

- Requiring user approval before sending emails.
- Preventing deletion of important files.
- Restricting financial transactions.
- Protecting confidential information.
- Refusing requests that violate organizational policies.

Safety constraints reduce operational risk and improve trustworthiness.

Step 11: Include Human Oversight

Many real-world systems include a human review stage before important actions are performed. Human oversight may be appropriate for activities such as:

- Approving contracts.
- Sending external communications.
- Processing financial transactions.
- Publishing information.

- Legal or medical recommendations.

Maintaining human involvement for high-impact decisions supports accountability.

Step 12: Build the Agent Workflow

The components developed in the previous steps are combined into a continuous workflow.

A simplified workflow generally follows this sequence:

1. Receive a goal or request.
2. Analyze the request.
3. Develop a plan.
4. Select appropriate tools.
5. Execute actions.
6. Evaluate results.
7. Update memory.
8. Determine whether additional work is required.
9. Return the completed result.

This execution cycle continues until the assigned objective has been completed or the agent determines that further progress requires human input.

Step 13: Test the System

Testing should occur before deployment.

Evaluation should include:

- Accuracy
- Reliability
- Response quality
- Tool selection
- Error handling
- Security
- Performance under unusual conditions

Testing with both expected and unexpected inputs helps identify weaknesses before users encounter them.

Step 14: Deploy the Agent

After testing, the agent can be deployed in an environment appropriate to its intended use.

Common deployment options include:

- Web applications

- Mobile applications
- Enterprise software
- Customer service platforms
- Internal business systems
- Workflow automation platforms
- Cloud-based services

Deployment also involves monitoring system performance and updating the agent as requirements evolve.

Step 15: Continuously Improve the Agent

Agentic AI systems continuous improvement which may involve:

- Refining instructions.
- Expanding tool capabilities.
- Updating knowledge sources.
- Improving planning strategies.
- Correcting errors
- Monitoring performance metrics.
- Enhancing security

Regular evaluation helps ensure that the agent remains effective throughout organizational needs, and user expectations changes.

Creating an agentic AI involves much more than connecting a language model to a user interface. A capable agent combines reasoning, memory, planning, external tools, retrieval of relevant information, safety mechanisms, and human oversight into a coordinated system that can pursue objectives while operating within defined boundaries. Successful development requires careful planning, rigorous testing, and ongoing refinement to ensure that the agent performs reliably, accurately, and responsibly in real world applications.

One lesson I have learned throughout my AI journey is that successful organizations rarely ask whether humans or AI should perform a task. Instead, they ask which parts of a task should be performed by humans and which parts should be performed by AI. That is a far more productive question. The future is unlikely to be defined by humans competing against machines. It is more likely to be defined by humans and machines working together, each contributing their respective strengths. Artificial Intelligence excels at speed, scale, consistency, and pattern recognition.

Human beings excel at judgment, wisdom, empathy, ethics, creativity, and leadership. The organizations that thrive will learn how to combine both. As executives, board members, entrepreneurs, and leaders, our responsibility is not simply to adopt Artificial Intelligence. Our responsibility is to adopt it wisely, responsibly, and strategically. Because ultimately, AI is not a technology strategy It is a business strategy, and organizations that understand that distinction will have a significant advantage in the years ahead. Although the terms are sometimes used interchangeably, a chatbot and an agentic AI are not the same type of system.

A chatbot is primarily designed to interact with users through conversation. Its main purpose is to receive a message, interpret it, and generate an appropriate response. A chatbot generally operates in a request-and-response pattern, meaning that it waits for the user to provide input before producing an output. Most chatbots do not independently plan future actions or pursue long-term goals unless those capabilities have been intentionally added.

An **agentic AI** works toward achieving objectives. Instead of responding only to individual prompts, an agentic AI can plan multiple steps, select and use external tools, evaluate the results of its actions, and decide what to do next based on its progress. Many agentic systems also include memory, planning mechanisms, and safeguards that allow them to perform more complex workflows. A chatbot can become part of an agentic AI system, but not every chatbot is an agent.

Building a Simple Chatbot Step by Step

The following example demonstrates how a basic chatbot can be designed before any programming begins.

Step 1: Define the Chatbot's Purpose

Every chatbot should have a clearly defined objective. For this example, suppose we are creating a chatbot for a public library. The chatbot will:

- Answer questions about library hours.
- Explain how to obtain a library card.
- Describe available services.
- Direct users to additional assistance when necessary.

Keeping the scope focused makes the chatbot easier to design and test.

Step 2: Identify the Intended Users

Next, determine who will interact with the chatbot.

Possible users include:

Library visitors
Students
Parents
Researchers
Community members

Understanding the audience influences the vocabulary, tone, and complexity of responses.

Step 3: Determine What Questions the Chatbot Should Answer

The chatbot should be designed around expected user questions.

Examples include:

- What are the library's operating hours?
- How do I apply for a library card?
- Can I reserve a study room?
- How long can I borrow books?
- Does the library offer digital resources?

These define the chatbot's knowledge domain.

Step 4: Organize the Knowledge

Before the chatbot can answer questions, accurate information must be collected. Example knowledge base:

Operating Hours

Monday through Friday: 9:00 a.m. to 6:00 p.m.

Saturday: 10:00 a.m. to 4:00 p.m.

Sunday: Closed.

Library Card

Residents may apply by presenting a government-issued photo identification and proof of address.

Book Loans

Books may be borrowed for three weeks and renewed if no other patron has requested them. In a real system, this information would come from the organization's official policies.

Step 5: Design the Conversation Flow

The chatbot follows a predictable sequence. Asks whether additional assistance is needed. This sequence forms the basic conversational workflow.

Step 6: Handle Greetings

The chatbot should recognize common greetings.

Example interaction:

User: Hello.

Chatbot: Hello. Welcome to the City Library. How may I assist you today?

Step 7: Answer Frequently Asked Questions

Example interaction:

User: What time does the library open?

Chatbot: The library opens at 9:00 a.m. Monday through Friday and at 10:00 a.m. on Saturdays.

Step 8: Handle Questions It Cannot Answer

No chatbot can answer every possible question.

When information is unavailable, the chatbot should respond honestly.

Example:

User: Can you recommend legal advice?

Chatbot: I do not have enough information to answer that question. Please contact the library staff for further assistance.

Step 9: Maintain Conversation Context

A chatbot should remember relevant information within the current conversation.

Example:

User: Tell me about library cards.

Chatbot: Library cards are available to residents with valid identification and proof of address.

User: How long are they valid?

Because the chatbot remembers the current topic, it understands that "they" refers to library cards.

Maintaining context improves the natural flow of conversation.

Step 10: Test the Chatbot

Testing should include expected and unexpected questions for example:

Expected: What are your hours? How do I get a library card?

Unexpected: Tell me today's weather. Solve this mathematics problem. Recommend a nearby restaurant.

Testing helps determine whether the chatbot remains within its intended scope.

Step 11: Improve the Chatbot

After testing, improvements may include:

- Adding additional questions and answers.
- Improving response clarity.
- Correcting inaccurate information.
- Expanding supported topics.
- Simplifying language for users.

Chatbots generally improve through repeated evaluation and refinement.

The Finished Chatbot Design

A simplified representation of the completed chatbot is shown below.

Unlike an agentic AI, this chatbot does not independently create plans, perform multiple tasks to reach a goal, or take actions outside the conversation. Its primary function is to understand user questions and provide accurate, relevant responses based on the information it has been given.

This distinction is why chatbots are often considered the foundation for more advanced AI systems. Many agentic AI applications begin with a conversational interface and then add capabilities such as planning, memory beyond the current conversation, tool use, and controlled autonomy to perform more sophisticated tasks.

Chapter 16

Prompt Engineering for Executives: How to Communicate with AI and Achieve Better Results

One of the most surprising lessons I learned during my exploration of Artificial Intelligence has been that success with AI has a lot to do with communication. When executives encounter advanced AI systems such as ChatGPT, Claude, Gemini, Grok, or similar large language models, the initial experience is often impressive. Within minutes, they witness the system generating reports, summarizing documents, answering questions, and producing content at remarkable speed. Yet the enthusiasm frequently fades. The responses begin to feel generic. The outputs become inconsistent, results might fail to meet expectations. Eventually, some users conclude that Artificial Intelligence is less capable than advertised. Whenever I hear this criticism, I ask a simple question: "What exactly did you ask the system to do?" In many cases, the answer reveals the underlying problem.

The issue is often not the intelligence of the system, but rather the quality of the instruction. There is a viral video of a skit where a patient returns to her asthma doctor to complain that her inhaler was not working. He asked her a few follow up questions and then asked her to demonstrate how she had been using the product and she sprayed the inhaler on her neck, as the doctor knowingly smiled.

The Communication Challenge

Human beings are exceptionally skilled communicators. We routinely understand one another through context, shared experiences, relationships, emotions, body language, tone of voice, and countless subtle cues that rarely appear in spoken or written language. A spouse can often anticipate what their partner is about to say before a sentence is completed. Longtime colleagues can identify concerns before they are expressed. Friends can recognize frustration, enthusiasm, or uncertainty from a brief conversation or facial expression. Artificial Intelligence does not possess these advantages. AI systems cannot access personal history, read facial expressions, interpret unspoken assumptions, or infer intentions with the same sophistication as human beings.

They can only work with the information provided to them. As a result, the quality of an AI-generated response is often closely linked to the quality of the instructions it receives. Clear instructions generally produce clearer outcomes and specific objectives usually generate more relevant results. Effective use of AI begins with effective communication.

Why Many Executives Struggle with AI

Many professionals expect AI to understand what they mean rather than what they actually communicate. AI responds to the instructions it receives, not to the intentions left unspoken. Consider a familiar leadership scenario. Imagine hiring an exceptionally talented university graduate who graduated near the top of their class and demonstrates outstanding analytical ability, learns quickly, and is eager to contribute. On their first day, you step into their office, say, "Prepare a report," and then walk away.

Most experienced leaders immediately recognize why this approach is unlikely to succeed. The request lacks the information needed to produce a meaningful result. What type of report should be prepared? Who is the intended audience? What business problem should it address? How detailed should it be? Should it be written as a formal report, a presentation, or an executive briefing? When is it due? Without answers to these questions, even the most capable employee would struggle to meet expectations. The issue would not be intelligence or motivation. It would be a lack of direction.

Ironically, this is exactly how many people interact with artificial intelligence. They provide brief prompts, limited context, and vague objectives while expecting precise, insightful, and highly relevant responses. When the outcome fails to meet expectations, they often conclude that the technology is inadequate, instead of recognizing that the instructions themselves were incomplete. In reality, AI frequently reflects the quality of the guidance it receives. Vague requests tend to produce vague responses. Unclear objectives often generate unfocused recommendations. When expectations are poorly defined, the results frequently miss the intended objective, not because the system lacks capability, but because it was never given enough information to understand what success looks like. One of the most valuable shifts leaders can make is to stop thinking of AI as a search engine and begin using it as a thinking partner.

A search engine is designed to retrieve information. It helps users locate facts, documents, and resources that already exist. Artificial intelligence can certainly perform those tasks, but its capabilities extend much further. When used thoughtfully, AI can help generate ideas, analyze complex situations, evaluate strategic alternatives, test assumptions, improve communication, and support better decision-making. Rather than simply retrieving information, it can participate in the thinking process itself.

This distinction is more important than it may first appear. Throughout my career in executive leadership, consulting, education, and organizational development, I have observed that meaningful decisions rarely emerge from collecting information alone. Facts are essential, but they are only the starting point. Strong decisions develop through discussion, exploration, reflection, and the disciplined examination of competing ideas.

Effective leaders ask thoughtful questions. They challenge assumptions, consider multiple perspectives, evaluate alternatives, and refine their thinking through dialogue. They recognize that better decisions are usually the product of better conversations. Modern AI can contribute meaningfully to that process. It cannot replace human judgment, experience, wisdom, or leadership. Those responsibilities remain firmly in human hands. However, AI can serve as a valuable intellectual resource that helps people broaden their thinking, identify additional possibilities, uncover blind spots, and evaluate ideas more efficiently.

When leaders approach AI in this way, it becomes much more than an information retrieval system. It becomes a practical thinking partner that supports structured reasoning and more informed decision-making. Many people assume that the greatest value of artificial intelligence lies in its ability to generate answers. While producing answers is certainly valuable, its greater contribution may be its ability to help people formulate better questions.

Although this distinction appears subtle, its implications are profound. Throughout history, transformative discoveries have often begun with a question rather than an answer. What if the Earth is not the center of the universe? What if diseases are caused by microscopic organisms? What if humans could achieve powered flight? What if information could be transmitted instantly across continents?

Each of these questions challenged accepted assumptions and opened entirely new avenues of discovery. The answers changed the world, but only because someone first asked a question that others had not considered. The same principle applies when working with artificial intelligence. The individuals who derive the greatest value from AI are not necessarily those with the deepest technical expertise. More often, they are people who know how to define problems clearly, provide meaningful context, question existing assumptions, and explore possibilities in a disciplined and systematic way.

In other words, they practice critical thinking before they ever begin prompting. Successful interactions with AI are rarely accidental. They are usually the result of clear communication. An effective prompt generally includes five essential elements:

1) **Clarity** requires defining the task or objective with precision so that the AI understands exactly what is being requested.
2) **Context** provides the relevant background information that

allows the response to reflect the situation accurately.

3) **Purpose** explains why the task matters and identifies the outcome that should be achieved.
4) **Constraints** establish important parameters such as audience, tone, format, length, deadlines, or other requirements that shape the response.
5) **Iteration** recognizes that high-quality results often emerge through refinement. By asking follow-up questions, clarifying expectations, and providing feedback, users can progressively improve both the quality and usefulness of the output.

Like any capable professional, AI performs best when expectations are communicated clearly. The quality of the results depends not only on the sophistication of the technology, but also on the clarity, completeness, and precision of the human communication guiding it. The better we communicate, the better AI can contribute. Rather than asking: "Write a strategy report." An executive might ask: "Prepare a five-page strategic analysis for the Board of Directors evaluating the opportunities and risks of adopting Artificial Intelligence within our organization over the next three years. Include and be detailed with financial considerations, operational implications, workforce impacts, and implementation recommendations."

The difference is substantial. The second instruction provides context, audience, scope, purpose, outcomes. The resulting response is likely to be significantly more useful.

The Human Advantage in the Age of AI

As Artificial Intelligence continues to become more capable, many people assume that human thinking will become less important. The opposite may be true.

AI can generate vast information and recommendations within seconds. However, it cannot on its own determine which objectives matter most or which values should guide decisions. Those aspects are fundamentally human. Judgment remains human. Wisdom remains human. Leadership remains human. The organizations that will benefit most from AI are unlikely to be those that merely adopt the newest technologies. They will be the organizations whose leaders learn how to ask better questions, define clearer objectives, and combine human insight with machine intelligence. The future belongs not simply to those who have access to Artificial Intelligence, but to those who know how to engage with it effectively.

Executive Takeaway

Prompt engineering is often described as the art and science of communicating with AI systems to achieve desired outcomes. For executives, however, it represents something even more important. It is the discipline of thinking clearly. The most effective AI users are rarely the most technical individuals in the room. They often communicate objectives with precision, with meaningful context, challenge assumptions, and remain curious enough to explore alternative perspectives. AI is not a replacement for human thinking. It is a tool that amplifies it.

The better we learn to communicate with these systems, the more valuable they become, not only as sources of information, but as partners in analysis, creativity, learning, and decision-making. Ultimately, the quality of the conversation determines the quality of the outcome. In the age of Artificial Intelligence, learning how to ask better questions may become one of the most important leadership skills of all.

The Future of Work and the Human Imperative

When I speak about AI, the conversation almost always begins with technology. People want to discuss the latest breakthroughs in large language models, machine learning, robotics, predictive analytics, autonomous systems. These are important discussions, and I enjoy them immensely. Yet over time, I have found myself becoming increasingly interested in a different question, one that I believe is ultimately far more important.

What will happen to people?

For all the attention devoted to technological advancement, every major innovation eventually arrives at the same destination, the workplaces, homes, schools, and communities. It arrives at the kitchen table where families discuss their future, their careers, their financial security, and the opportunities available to the next generation. Technology may originate in laboratories, universities, and corporate boardrooms, but its true impact is measured by how it affects human lives.

During my studies at Oxford, I had the privilege of learning alongside executives, entrepreneurs, and professionals from many different countries and industries. Our discussions often focused on some of the most advanced technological developments of our time. Yet beneath the conversations about algorithms, data, and innovation, I repeatedly encountered a deeper concern. Regardless of industry or geography, people were asking variations of the same question. How do we ensure that technological progress benefits society without leaving people behind?

That question has remained with me ever since. History offers valuable perspective. Every major technological revolution has created both opportunities and disruptions. The Industrial Revolution transformed agriculture, manufacturing, transportation, and commerce, contributing to unprecedented economic growth while also fundamentally reshaping labor markets and communities. The digital revolution connected billions of people, transformed communication, and created entirely new industries, yet it also displaced established business models and changed the nature of work.

Artificial Intelligence is likely to follow a similar trajectory. It has the potential to increase productivity, accelerate innovation, improve decision-making, and solve complex problems that have challenged organizations for decades. At the same time, it will inevitably alter how work is performed, how skills are valued, and how organizations are structured. Acknowledging this reality is not an argument against technological progress. Rather, it is an argument for thoughtful leadership.

One concept that emerged from my reflections is what I have come to think of as a need for organizations and leaders to adopt a philosophy of "Humane Resources." Most organizations have Human Resources departments, and the term has become so familiar that few people pause to consider its meaning. Yet as I reflected on the future of work, I began to feel that the traditional language did not fully capture the challenge ahead. Human Resources focuses on managing people as organizational assets. Humane Resources would focus on developing, supporting, and caring for people as human beings with intentionality towards reskilling and upskilling in this technological evolution.

The distinction may appear subtle, but I believe it will become increasingly important in the age of Artificial Intelligence. As AI systems become more capable, organizations will face difficult decisions. Certain tasks will become automated. Some roles will evolve significantly. Other positions may disappear altogether. At the same time, entirely new professions and opportunities will emerge. This pattern has accompanied every major technological shift throughout history. The critical question is not whether change will occur. Change is inevitable. The real question is how leaders choose to respond.

When organizations discover that AI can automate a good portion of a department's workload, the immediate temptation may be to focus solely on efficiency gains and workforce reductions. While efficiency is an important business objective, it should not be the only consideration. Forward-thinking leaders will ask a different set of questions. How can employees be prepared for new roles? What skills will become more valuable in the future? How can technology be used to augment human capabilities rather than simply replace them?

These are fundamentally leadership questions rather than technology questions. In many respects, I believe the organizations that thrive in the coming decades will not necessarily be those with access to the most advanced technologies. They will be the organizations that become exceptional at learning. The pace of technological change is accelerating. Skills that were highly valuable only a few years ago may become less relevant as industries evolve and new capabilities emerge. In such an environment, education can no longer be viewed as a phase of life that ends with graduation. Learning must become a continuous process.

This requires organizations to treat upskilling and reskilling as strategic priorities rather than occasional initiatives. It requires leaders to create cultures that encourage curiosity, adaptability, and lifelong learning. Most importantly, it requires a recognition that the workforce of the future will be defined less by what people know today and more by their capacity to learn tomorrow. This shift may ultimately become one of the most significant consequences of Artificial Intelligence.

For much of modern economic history, professional success was closely linked to specialized knowledge. Individuals accumulated expertise over decades, and that expertise became a source of competitive advantage. Today, information is increasingly abundant and accessible. AI systems can retrieve, summarize, and organize vast amounts of information in seconds.

As information becomes easier to access, other qualities become more valuable. Judgment becomes more valuable. Wisdom becomes more valuable. Leadership becomes more valuable. The ability to navigate ambiguity, build trust, manage complexity, inspire teams, and make decisions under conditions of uncertainty becomes more valuable.

These are deeply human capabilities. They are difficult to automate because they depend on context, experience, values, emotional intelligence, and ethical reasoning.

Ironically, as machines become more capable, the uniquely human dimensions of work may become even more important.

This observation has significant implications for workforce development. Technical literacy, digital fluency, and data literacy will undoubtedly remain essential. However, organizations must also invest in communication skills, creativity, collaboration, critical thinking, empathy, and ethical decision-making. These capabilities are not secondary to technological competence; they are increasingly complementary to it.

The leaders who succeed in the age of AI will therefore need to understand more than technology. They will need to understand people. This is why I often say that every AI strategy is ultimately a people strategy. Organizations frequently describe AI initiatives as technology transformations, yet the success or failure of these efforts rarely depends on technology alone. Success depends on whether employees trust the process, understand the purpose, acquire the necessary skills, and feel included in the journey. Every digital transformation is, at its core, a human transformation.

Preparing people for change before disruption occurs is therefore one of the most important responsibilities of leadership. Organizations that wait until jobs have been eliminated before discussing workforce development have already fallen behind. By contrast, organizations that begin investing in learning, adaptability, and workforce readiness today will be far better positioned to navigate the uncertainties of tomorrow.

Consequently, the discussions about Artificial Intelligence cannot focus exclusively on algorithms, software platforms, and technical performance. They must also address education, workforce development, social mobility, opportunity, and human dignity.

One of the dangers in contemporary AI discussions is that they can become conversations among technologists speaking primarily to other technologists. Yet many of the people who will be most affected by AI are not engineers, data scientists, or software developers. They are teachers, healthcare professionals, small business owners, retirees, community leaders, parents, and citizens seeking to understand how these technologies may influence their lives and livelihoods.

Their concerns are legitimate. They deserve attention. The future of Artificial Intelligence should not be measured solely by efficiency gains, productivity improvements, or financial returns. Those outcomes are important, but they are not sufficient on their own. The broader objective should be to ensure that technological advancement contributes to human flourishing. Innovation should create opportunity, expand human potential, and improve quality of life.

Ultimately, the most important AI strategy is not a technology strategy at all. It is a leadership strategy grounded in responsibility, adaptability, and a commitment to people. The future will undoubtedly bring extraordinary technological capabilities. The question is whether we will have the wisdom to deploy them in ways that strengthen individuals, organizations, and society as a whole. That is the future I hope we build. On in which technological progress and human progress advance together, and where innovation is guided not only by artificial intelligence, but also by humanity.

One of the problems in AI discussions is that they often become conversations among technologists talking to other technologists. Yet many of the people most affected by AI are not engineers. They are retirees, grandparents, teachers, small business owners, community leaders, and ordinary citizens who are trying to understand whether this new technology is something to embrace, fear, or simply tolerate.

Chapter 18

Agentic AI and the Next Frontier - The Future Is Arriving Faster Than We Think

One of the most insightful conversations I have had about artificial intelligence did not involve a professor, technology executive, or researcher. It involved my mother-in-law. She belongs to a generation that has witnessed some of the most significant technological transformations in modern history. She has seen handwritten letters replaced by email, rotary telephones evolve into smartphones, and personal computers become indispensable tools of everyday life. She has lived through the rise of the internet, social media, online banking, video conferencing, and countless innovations that once seemed unimaginable.

What makes her perspective particularly valuable is that she approaches technology with a balance of curiosity and caution. She is an accomplished intelligent retiree still actively positively impacting her community, eager to learn, willing to explore new possibilities, and appreciative of the benefits technology. At the same time, she asks thoughtful and important questions. "What happens if the technology makes a mistake?" "What happens if people become overly dependent on it?" "What happens if it falls into the wrong hands?" "What happens when innovation moves faster than society's ability to govern or manage it?"

Over time, I have come to appreciate these questions more deeply. In fact, I believe they are precisely the questions we should be asking. Too often, discussions about artificial intelligence become polarized. On one side are those who view AI as the solution to nearly every challenge facing humanity. On the other are those who see it as too risky or even an existential threat. Reality is rarely found at extreme. The future is seldom as transformative as enthusiasts predict in the short term, nor as catastrophic as critics fear.

More often, technological change unfolds in complex and new ways, creating extraordinary opportunities while introducing new challenges that require careful judgment and responsible leadership. This perspective is especially important as we enter what many experts believe will be the next major phase of artificial intelligence. Agentic AI.

One of the limitations of many public discussions about AI is that they remain conversations among technologists speaking to other technologists. The people who will be most affected by AI are not necessarily engineers or data scientists.

They are teachers, healthcare professionals, small-business owners, public servants, community leaders, retirees, and citizens seeking to understand whether this technology should be embraced, feared, or simply understood. Most AI systems people interact with today are fundamentally reactive. A user asks a question, and the system provides an answer. A prompt is entered, and the system generates a response. The interaction begins and ends with human instruction. Agentic AI changes that relationship.

Rather than simply responding to requests, agentic systems are designed to pursue objectives. They can plan, reason across multiple steps, evaluate alternatives, adapt to changing conditions, and coordinate actions across different tools and environments.

Consider the task of organizing an international conference. A traditional AI system might help draft invitations, suggest venues, or generate a project plan. An agentic system could go much further. It could compare venue options, identify potential speakers, help to coordinate schedules, draft correspondence, monitor registrations, track budgets, and provide ongoing status updates throughout the process. In other words, it moves beyond assisting with individual tasks and begins helping manage entire workflows.

This shift is significant because execution, not information, is where value is ultimately created. Strategies only matter when they are implemented. Plans only matter when they produce results. The movement from content generation to autonomous execution represents one of the most consequential developments in the evolution of artificial intelligence.

The Rise of Autonomous Workflows

Although the concept may sound futuristic, its foundations are already emerging. Organizations across industries are experimenting with AI agents capable of scheduling meetings, managing customer inquiries, conducting research, monitoring regulatory requirements, reviewing contracts, analyzing financial reports, and coordinating operational processes.

More advanced environments are beginning to explore systems in which multiple AI agents work together. Imagine a marketing agent collaborating with a sales agent. A finance agent coordinating with an operations agent. A customer-service agent communicating with a planning agent. Each performs a specialized function while contributing to a shared objective. What once appeared to be science fiction is gradually becoming operational reality. The implications for productivity, efficiency, and organizational design are profound. Yet agentic systems represent only one dimension of a much larger transformation.

When AI Meets the Physical World

For many years, robotics remained largely confined to industrial manufacturing environments. Today, however, robots are becoming more mobile, more intelligent, and increasingly capable of operating in dynamic real-world conditions. At the same time, drones are reshaping industries ranging from agriculture to emergency response. Farmers use them to monitor crops, identify irrigation challenges, and optimize resource allocation. Emergency responders deploy them during natural disasters to assess damage and locate survivors. Infrastructure operators use them to inspect assets, while healthcare organizations are exploring their use in delivering medical supplies to remote regions.

Advances in imaging technologies are accelerating this transformation even further. The future of AI is therefore not confined to software applications running on computers. Increasingly, artificial intelligence is becoming the intelligence layer that connects digital systems, sensors, robotics, drones, satellites, and human decision-making processes into integrated ecosystems. The next wave of innovation will emerge not from AI alone, but from the convergence of AI with the physical world.

The Great Convergence

Despite rapid advances, today's AI landscape remains highly fragmented. Professionals often use one platform for writing, another for image creation, another for video generation, another for voice synthesis, another for analytics, another for workflow automation, and yet another for research. Valuable time is lost navigating between systems, formats, and disconnected processes. This fragmentation creates friction. Yet it also creates one of the greatest business opportunities of the coming decade.

The organizations that successfully unify these capabilities into seamless user experiences may unlock extraordinary value. Imagine issuing a single instruction that generates a strategic plan, creates a presentation, produces supporting graphics, develops video content, drafts communications, launches a marketing campaign, and continuously monitors performance. The trajectory is increasingly clear. The long-term direction of AI is moving toward integration as well as autonomous execution across multiple domains.

The Enduring Importance of Human Judgment

Whenever people ask whether artificial intelligence is becoming too powerful, I often think again of my mother-in-law. Her perspective reflects what may be the most appropriate response to this technological moment: cautious optimism. We should not fear technology simply because it is new. Nor should we embrace it uncritically simply because it is impressive. Instead, we should seek to understand it, evaluate its risks honestly, and pursue its opportunities responsibly.

Technology is, and always has been, a tool. Its impact depends largely on the wisdom, values, and judgment of those who design it, deploy it, and govern its use. As powerful as agentic AI, robotics, drones, and autonomous systems may become, they do not eliminate the need for human judgment. If anything, they increase it. The more capable our tools become, the more important it becomes to determine how those tools should be used, where boundaries should be established, and which decisions should remain under human control. That responsibility remains uniquely human. And it is likely to remain so for a very long time.

The future of artificial intelligence is not merely a story about technology. It is a story about people, institutions, values, and choices. The systems we build will shape the world we inhabit, but the decisions we make will determine whether that world becomes more prosperous, more equitable, and more resilient. The next chapter of AI is already being written.

Chapter 19

What Comes Next? Artificial General Intelligence (AGI)

Artificial General Intelligence (AGI) is a form of artificial intelligence that can understand, learn, reason, adapt, and solve problems across many different domains at a level comparable to a human being. Unlike today's AI systems, which are designed for specific tasks, AGI would be able to transfer knowledge from one area to another, learn new skills without being specifically programmed, and perform a wide variety of intellectual tasks.

Simple Definition

Think of AGI as an AI that can do anything a reasonably educated human can do intellectually, rather than being limited to a single specialty.

For example:

- A calculator can perform mathematics.
- A language model can write essays and answer questions.
- An image generator can create artwork.
- A chess engine can play chess.

Each of these systems is excellent at its particular task.

An AGI, however, could potentially:

- Write a business plan.
- Learn a new language.
- Diagnose a technical problem.
- Teach a child mathematics.
- Design a bridge.
- Compose music.
- Analyze a legal contract.
- Learn how to perform a task it has never seen before.

all using the same underlying intelligence.

Understanding AGI Through Examples

Today's AI systems are highly specialized. A calculator performs mathematical computations, a language model generates and analyzes text, an image-generation system creates visual content, and a chess engine excels at strategic gameplay. Each system demonstrates remarkable capability within its designated domain but remains limited to that specific function.

An AGI system, by contrast, would be capable of performing a broad spectrum of tasks using a unified intelligence. Such a system could potentially:

- Write a business plan.
- Learn a new language.
- Diagnose technical problems.
- Teach mathematics to a child.
- Design engineering solutions.
- Compose music.

- Analyze legal contracts.
- Learn entirely new tasks without prior programming.

The defining characteristic of AGI is not excellence in one area but competence across many areas through general reasoning and learning capabilities.

AGI and Contemporary Artificial Intelligence

Today's AI (Narrow AI)	Artificial General Intelligence (AGI)
Designed for specific tasks	Capable of performing many different tasks
Requires specialized training	Learns and adapts across domains
Limited transfer of knowledge	Applies knowledge broadly

Excels within predefined boundaries	Operates across diverse contexts
Examples include ChatGPT, Siri, image generators, and chess engines	Does not currently exist

While modern AI systems may appear highly intelligent, they remain specialized tools rather than genuinely general intelligences.

Core Characteristics of AGI

1. Reasoning

A defining feature of AGI would be the ability to reason effectively in unfamiliar situations. For example, if presented with a completely new board game, a narrow AI system would typically require extensive training data and numerous examples. An AGI, however, could read the rules, understand the objectives, develop strategies, and improve through experience in a manner similar to a human learner.

2. Continuous Learning

AGI would be capable of learning continuously from experience rather than relying solely on predefined training. Human beings routinely transfer knowledge from one activity to another. A person who learns concepts such as balance, motion, and coordination while riding a bicycle can apply those principles when learning to ride a motorcycle, skateboard, or skis. AGI would similarly transfer insights and knowledge across different tasks and environments.

3. Adaptability

Another key attribute of AGI would be its capacity to adapt to unfamiliar circumstances. Consider an organization that requires assistance in inventory management, accounting, customer analysis, and coordination optimization. Current AI systems often require separate specialized tools for each function. An AGI could potentially learn and perform all these tasks while adapting its approach as organizational needs evolve.

4. Common-Sense Understanding

Humans rely heavily on common-sense reasoning to interpret everyday situations. For instance, when hearing the statement, "The ice cream melted because it was left in the sun," people immediately understand the causal relationship between heat, sunlight, and melting. AGI would ideally possess similar forms of practical understanding, allowing it to interpret information beyond literal patterns and statistical associations.

5. Planning and Goal Achievement

AGI would likely be capable of setting objectives, developing strategies, and executing long-term plans. For example, if tasked with reducing traffic congestion in Lagos, an AGI might:

- Analyze traffic-flow data.

- Study demographic and population trends.
- Evaluate public transportation systems.
- Model infrastructure improvements.
- Recommend policy interventions.
- Monitor outcomes and adjust strategies over time.

Such planning would involve integrating knowledge from multiple disciplines while adapting to changing conditions.

A Practical Illustration of AGI

To understand the potential scope of AGI, consider a disaster-response scenario following severe flooding.

Situation Assessment

The AGI reviews:

- Weather forecasts.
- Satellite imagery.

- Emergency service reports.
- Hospital capacity data.

Strategic Planning

The system identifies:

- Accessible transportation routes.
- Locations for emergency shelters.
- Priority areas for resource allocation.
- Efficient supply-distribution strategies.

Communication

The AGI:

- Drafts public advisories.
- Translates information into multiple languages.
- Responds to public inquiries.
- Coordinates communication among agencies.

Continuous Adaptation

As conditions change, the system updates recommendations, revises operational plans, and learns from new information.

While current disaster-management efforts typically require multiple specialized technologies and teams, AGI could potentially integrate these functions within a single adaptive system.

Potential Benefits of AGI

If developed responsibly and aligned with human interests, AGI could contribute significantly to societal progress.

Healthcare

AGI could:

- Accelerate medical research.
- Support clinical diagnosis.
- Assist in drug discovery.
- Improve treatment planning.

Education

Potential applications include:

- Personalized tutoring.
- Adaptive learning pathways.
- Real-time translation.
- Expanded access to educational resources.

Scientific Research

AGI could assist researchers by:

- Discovering new materials.
- Advancing renewable-energy technologies.
- Solving complex mathematical problems.
- Accelerating scientific innovation.

Government and Public Services

Applications may include:

- Transportation optimization.
- Enhanced disaster response.
- Public service improvement.
- Evidence-based policymaking.

Business and Industry

Organizations could benefit through:

- Strategic planning.
- Operational optimization.
- Market analysis.
- Process improvement.
- Data-driven decision-making.

Risks and Challenges

Despite its potential benefits, AGI also presents significant challenges and risks.

Alignment Challenges

One of the most widely discussed concerns is ensuring that AGI systems pursue objectives that remain consistent with human values and intentions. Misaligned goals could produce unintended or harmful outcomes.

Economic Disruption

AGI could automate many cognitive and knowledge-based occupations, potentially transforming labor markets and creating substantial economic and social challenges.

Security Concerns

Highly capable systems may be vulnerable to misuse by malicious actors, creating new risks in cybersecurity, misinformation, and other sensitive areas.

Concentration of Power

Organizations or governments with access to advanced AGI technologies could acquire disproportionate economic, political, or strategic influence.

Does AGI Exist Today? No. AGI has not yet been achieved.

Current AI systems developed by leading organizations, including OpenAI, Google DeepMind, and Anthropic, are generally categorized as advanced forms of narrow AI rather than AGI. Although these systems demonstrate impressive capabilities across many tasks, they continue to face limitations in areas such as:

- Long-term autonomous reasoning.
- Robust common-sense understanding.
- Self-directed learning.
- Human-level adaptability across all domains.

Consequently, most researchers and industry experts do not consider existing AI systems to be examples of true AGI.

A Business and Leadership Perspective

Consider a scenario in which the President of the American Women's Club Lagos seeks to increase membership by 50 percent, improve fundraising outcomes, strengthen community engagement, and develop a sustainable five-year strategic plan.

Contemporary AI tools can support individual aspects of this objective by generating marketing ideas, creating budget templates, or analyzing survey results.

A true AGI, however, could potentially take a far more integrated approach by:

- Analyzing membership trends.
- Studying comparable organizations globally.
- Developing financial projections.
- Designing outreach strategies.
- Identifying strategic partnerships.
- Monitoring implementation.
- Continuously adapting recommendations as circumstances evolve.

In this sense, AGI would function less as a specialized tool and more as a highly capable strategic collaborator. Artificial General Intelligence represents the vision of a machine capable of human-level, flexible intelligence across a broad range of intellectual activities. Unlike today's specialized AI systems, AGI would be able to learn, reason, adapt, plan, and transfer knowledge between domains in ways that resemble human cognitive capabilities.

Although AGI remains theoretical, it is widely regarded as one of the most transformative technological possibilities of the twenty-first century. Its development could unlock significant advances in science, education, healthcare, business, and public service, while simultaneously raising profound ethical, economic, and governance challenges that will require careful global consideration.

Chapter 20

The CEO's AI Playbook

Turning Curiosity into Competitive Advantage

By the time most executives begin thinking seriously about Artificial Intelligence, they have already been exposed to years of discussion, speculation, and increasingly bold predictions. They have attended conferences where AI was described as the defining technology of the century.

They have listened to consultants explain how industries are being transformed. They have watched competitors announce new AI initiatives and read headlines suggesting that organizations that fail to adopt AI risk being left behind. Yet despite the abundance of commentary around Artificial Intelligence, many leaders still find themselves confronting a surprisingly simple question when they return to their offices: What exactly should we do next?

Throughout my studies at Oxford and in conversations from different parts of the world, I was struck by how consistent the concerns were. Regardless of industry or geography, the discussion inevitably moved beyond the technology itself and toward implementation.

Leaders wanted practical answers rather than theoretical explanations. They wanted to know where to begin, how much to invest, whether to build their own solutions or purchase existing ones, how to evaluate vendors, how to measure success, and how to govern AI responsibly. These are not primarily technical questions. They are leadership questions, and leadership is where every successful AI transformation begins.

One of the misconceptions surrounding Artificial Intelligence is the belief that it is technology project. With this, organizations often respond to AI in the same way they have historically responded to other technological developments and assign responsibility to the information technology department, approve a budget, evaluate software vendors, and expect innovation to follow. Months later, many of these organizations discover that little has changed. New tools may have been deployed, but productivity remains largely unchanged, decision-making has barely seen improvement and anticipated transformation may have failed to materialize.

The reason could be straightforward. Technology alone has never transformed organizations. Organizations are transformed by people, leadership, culture, and processes. Technology can accelerate change, enable new capabilities, and create opportunities for innovation, but it cannot provide strategic direction or organizational purpose. The companies, universities, government agencies, and nonprofit institutions that are generating meaningful value from AI are not necessarily those with the most advanced technology. They are the organizations that understand where they create value and have carefully examined how AI can strengthen that value.

For this reason, the first question leaders should ask is not which AI platform they should purchase. A far more important question is where time, effort, and resources are currently being lost within the organization. Every institution has bottlenecks. Employees spend hours searching for information that should be easily accessible. Managers devote significant amounts of time to preparing reports that are largely repetitive.

Customer service teams answer the same questions countless times. Legal professionals review large volumes of contracts and documentation. Finance departments perform routine reconciliations and compliance checks. Marketing teams spend weeks producing content that may only be used once. These activities are not usually the most visible parts of an organization, but they often represent some of the greatest opportunities for AI-driven improvement.

Importantly, the objective should not be replacement. Public discussion about AI frequently focuses on automation and job displacement, but the most successful implementations are often those that enhance human capability rather than eliminate it. Organizations achieve the greatest returns when AI helps employees work more effectively, make better decisions, access information more quickly, and devote more time to higher-value activities. In practice, the most successful AI initiatives rarely begin with ambitious promises of revolutionary change. They begin with practical improvements that solve real business problems and generate measurable results.

During discussions at Oxford about technology adoption and organizational transformation, another recurring theme emerged. Institutions frequently overestimate what can be accomplished in the short term and underestimate what can be achieved over the longer term. This pattern is especially visible in AI adoption. Once leadership teams become convinced that AI is important, there is often a temptation to launch multiple initiatives simultaneously.

Every department wants its own AI strategy, every project acquires an AI component, and every strategic conversation begins to revolve around emerging technologies. While enthusiasm is valuable, it can quickly outpace execution.

Organizations that achieve lasting success with AI typically adopt a more disciplined approach. Rather than attempting to transform everything at once, they identify a small number of high-impact opportunities and execute them exceptionally well. Early successes create confidence throughout the organization. Confidence encourages broader adoption. Broader adoption generates momentum. Over time, that momentum becomes genuine transformation. In many cases, a single successful project can accomplish more than a dozen poorly executed initiatives.

One of the questions I am asked most frequently is whether organizations should build their own AI solutions or purchase existing technologies. The answer depends largely on the strategic importance of the capability being developed and the resources available to the organization. There is often a tendency among executives to assume that building proprietary systems will create a competitive advantage. While that may occasionally be true, the reality is that building and maintaining sophisticated AI systems requires significant expertise, substantial investment, ongoing governance, and continuous adaptation as technologies evolve.

For most organizations, purchasing proven solutions and integrating them effectively will provide greater value and lower risk than attempting to build complex systems from scratch. Unless AI itself is central to the organization's competitive differentiation, buying is often the more practical and economically sound choice.

Equally important is the selection of technology partners. The AI marketplace is evolving rapidly, and vendors frequently compete through impressive demonstrations and ambitious claims. Effective leaders learn to look beyond demonstrations and focus on business outcomes. The critical question is not whether a technology appears impressive, but whether it solves a meaningful organizational problem. Issues such as data security, regulatory compliance, scalability, transparency, financial stability, and long-term support should receive as much attention as technical performance. Technology should serve strategy, not the other way around.

Perhaps the most important lesson emerging from the AI era is that successful implementation depends far more on organizational readiness than on technological sophistication. Many leaders ask who should own AI within the organization. In reality, no single individual can own a transformation of this magnitude. The chief executive provides strategic direction, technology leaders oversee implementation, legal and compliance teams manage risk, human resources prepares the workforce, business units identify opportunities, and the board provides governance and oversight. AI cannot operate as an isolated initiative managed on the margins of the organization. It must become integrated into how the institution thinks, plans, operates, and creates value.

As AI capabilities become increasingly powerful, governance assumes greater importance. Every transformative technology in history has required responsible oversight, and AI is no exception. Organizations must establish clear frameworks governing privacy, security, accountability, transparency, ethical use, and human oversight. Employees need guidance regarding appropriate use. Customers need confidence that their data is protected. Stakeholders need assurance that AI is being deployed responsibly. In an era where trust is becoming a significant competitive advantage, responsible governance is no longer optional. It is a strategic necessity.

This responsibility extends directly to the boardroom. Artificial Intelligence is no longer merely an operational issue delegated to management. It is increasingly a matter of strategic oversight. Board members do not need to become technical experts, but they do need sufficient understanding to ask informed questions about risk, governance, workforce readiness, data protection, performance measurement, and accountability. The quality of those questions may ultimately determine the quality of the organization's AI strategy.

When discussing AI with executives, the conversation inevitably returns to return on investment. Leaders want to know how success should be measured. The answer is no different from the evaluation of any other strategic initiative. Organizations should assess whether productivity has improved, whether costs have been reduced, whether customer experiences have improved, whether decision-making has become more effective, whether risks have been mitigated, and whether new opportunities for growth have emerged. Artificial Intelligence should not be judged by its sophistication or novelty. It should be judged by the value it creates.

Reflecting on my Oxford experience and on the many conversations that have shaped my understanding of Artificial Intelligence, one conclusion stands out above all others. The leaders succeeding in the AI era are not necessarily those with the deepest technical expertise. More often, they are the individuals who remain curious, adaptable, and willing to learn. They recognize that technology will continue to evolve, platforms will change, and capabilities will expand, but the fundamental principles of leadership remain remarkably constant. Vision, judgment, integrity, accountability, and the ability to guide people through uncertainty remain as important today as they have ever been.

The organizations that thrive in the age of Artificial Intelligence will not do so because they possess the most sophisticated algorithms. They will thrive because their leaders understand how to combine technological capability with human wisdom. Ultimately, the future belongs not to organizations that merely adopt AI, but to those that learn how to use it thoughtfully, responsibly, and strategically in pursuit of their broader mission.

This chapter marks the transition from understanding Artificial Intelligence to leading with it. Yet another question remains. If today's AI systems are already reshaping industries, how might tomorrow's systems transform society? As advances in autonomous agents, robotics, multimodal systems, and Artificial General Intelligence continue to accelerate, leaders must prepare not only for the present wave of innovation but also for what comes next. That future is the subject of our final chapter.

The CEO's AI Playbook - Turning Curiosity into Competitive Advantage

The question facing organizations today is no longer whether they should engage with artificial intelligence. That debate has largely passed. The more important question is how leaders can adopt AI responsibly, strategically, and in ways that create lasting value. Throughout this book, we have explored the evolution of artificial intelligence, examined its capabilities and limitations, discussed its opportunities and risks, and considered the profound ways it is reshaping business and society. Those conversations provide essential context, but they ultimately lead to a more practical question. What happens next?

This is the chapter for Monday morning. It begins after the conference has ended, the keynote speaker has left the stage, and the consultants have packed their laptops. It is the chapter where ideas must become decisions and strategy must become execution. Every executive eventually arrives at the same question, "What should we do now?"

During my studies and discussions at Oxford, I was struck by how consistently leaders asked the same questions regardless of industry, geography, or organizational size. Whether I was speaking with business executives, policymakers, entrepreneurs, nonprofit leaders, or academics, the conversations repeatedly returned to a familiar set of concerns. Where should we begin? How much should we invest? Should we build our own AI capabilities or purchase established solutions? How do we select trustworthy vendors? How should boards oversee AI? How do we measure success? Just as importantly, what mistakes should we avoid?

Although these questions involve technology, they are not fundamentally technical questions. They are leadership questions, and leadership is where every successful AI transformation begins.

Every organization has recurring points of friction that consume valuable time and resources. Employees spend countless hours searching for information that already exists. Managers devote days to preparing reports that require repetitive analysis.

Customer service representatives answer the same questions hundreds of times each week. Legal departments review thousands of contracts. Finance teams perform routine reconciliations, while marketing departments invest weeks developing campaigns that could often be accelerated without sacrificing quality.

These activities are not signs of poor performance. They are simply examples of work that is repetitive, time-consuming, and well suited for intelligent automation and decision support. The goal, however, is not to replace people. It is to improve how people work. The most successful AI initiatives rarely begin with ambitious promises of complete organizational transformation. Instead, they begin by solving a meaningful business problem that produces measurable value. Saving hundreds of hours of repetitive work or improving customer response times often creates far more momentum than a highly publicized AI initiative that fails to deliver tangible results.

Organizations that achieve lasting success with AI focus first on solving real business challenges. They understand that technology is valuable only when it advances the organization's mission and strategy. Another lesson reinforced throughout my conversations at Oxford was the importance of disciplined experimentation.

When organizations first embrace AI, enthusiasm can quickly outpace execution. Suddenly every initiative becomes an AI initiative. Every department requests funding. Every strategic presentation includes AI, and every meeting turns into a discussion about new possibilities. While excitement is understandable, it can easily become a distraction.

Organizations that achieve sustainable success typically resist the temptation to pursue dozens of initiatives simultaneously. Instead, they identify one or two opportunities where AI can deliver measurable business value, execute those projects exceptionally well, and use those early successes to build confidence across the organization.

Small victories create credibility. Credibility builds trust. Trust encourages broader adoption. Over time, that momentum becomes transformation. The organizations leading the AI era did not begin by transforming everything. They began by transforming something important and doing it well.

Chapter 22

Building the AI Organization - Who Should Do What?

The greatest challenge in AI adoption is rarely the technology itself. More often, it is the organization. Technology is only one component of successful AI transformation. Leadership, governance, culture, data, operations, workforce readiness, and strategic alignment are equally important. Organizations that view AI solely through a technological lens frequently struggle because they fail to recognize that they are not simply introducing a new tool, but a new capability that will influence how decisions are made, how work is performed, and how value is created. In that sense, AI is not merely a technology project. It is an organizational transformation initiative.

This distinction becomes important when leaders begin asking practical questions about implementation.

The instinctive response in many organizations is to ask which platform should be purchased, which vendor should be selected, or which software should be deployed first. Yet the more important question is often overlooked. Who will lead the effort? One main reasons AI initiatives lose momentum is that responsibility is never clearly established. Technology teams assume the business will define priorities.

Successful organizations approach the challenge differently. They begin by establishing ownership at the highest level of leadership. This does not mean that the Chief Executive Officer must become an AI specialist or spend time studying machine learning algorithms. It does mean that the CEO must become the visible sponsor of the transformation. Serious organizational change requires executive commitment where leaders consistently communicate the importance of AI, connect it to the organization's mission, and demonstrate personal engagement with the transformation. Then the rest of the organization pays attention. When leadership treats AI as a side project delegated to a technical team, employees tend to do the same.

The CEO's role is therefore not to build the technology but to provide direction. Effective leaders ensure that AI initiatives support broader organizational objectives rather than becoming isolated experiments. They help determine where AI can create value, where investment should be focused, and how success will be measured. Most importantly, they establish the conditions under which innovation can occur while maintaining accountability and trust.

Once executive commitment becomes established, and AI capabilities become embedded across multiple functions, it is important for someone to oversee the broader roadmap. Some organizations have responded by appointing a Chief AI Officer, while others assign responsibility to an existing executive. The title matters less than the mandate.

Someone must have visibility across departments, understand strategic priorities, coordinate resources, and ensure that individual projects contribute to a coherent organizational singular vision. Without this coordination, AI initiatives often emerge independently within different departments, creating inefficiency, and glaring inconsistent outcomes.

Equally important is the role of data leadership. During my Oxford studies, one principle surfaced repeatedly in discussions about AI success and failure. Organizations often underestimate the importance of data readiness. There is a tendency to focus on sophisticated models and emerging technologies while overlooking the quality of the information that powers them. Yet AI systems are only as effective as the data on which they rely. Poor-quality data, fragmented systems, inconsistent governance, and inadequate infrastructure can undermine even the most ambitious AI initiatives. Whether the responsibility sits with a Chief Data Officer or another senior executive, organizations must ensure that their information assets are accurate, accessible, secure, and aligned with business objectives. Investing in AI without investing in data readiness is rather like purchasing a sophisticated aircraft before constructing a runway. The technology may be impressive, but it cannot achieve its intended purpose without the necessary foundation.

Technology teams, of course, remain essential. Their role, however, extends far beyond software deployment. As AI becomes integrated into core business processes, technology leaders become stewards of an increasingly complex digital ecosystem. They evaluate platforms, manage integrations, protect cybersecurity, ensure operational resilience, monitor performance, and maintain scalability. Their responsibility is not simply to implement systems but to ensure that those systems remain reliable, secure, and capable of supporting the organization's evolving needs. In many respects, they provide the infrastructure that allows innovation to flourish safely.

Perhaps the most underestimated function in AI transformation is Human Resources, or what I increasingly think of as Humane Resources. Technology discussions often focus on systems and capabilities, yet the human dimension is frequently where success or failure is ultimately determined. Employees hear discussions about automation, efficiency, and artificial intelligence, and it is entirely natural for some to wonder what these changes mean for their own future. When organizations fail to address these concerns openly, uncertainty can quickly develop into anxiety, and anxiety often leads to resistance.

The most successful organizations recognize that workforce preparation is not a secondary consideration; it is a strategic priority. Employees should not experience transformation as something that happens to them. They should experience it as something they are prepared for and invited to participate in.

This requires clear communication, continuous learning opportunities, skills development, leadership training, and thoughtful workforce planning. Organizations that invest in helping their people adapt to AI often discover that their workforce becomes some of the strongest advocates for change. CEOs who neglect this frequently encounter resistance that no technology solution can overcome. Legal, compliance, and risk management-functions also play a vital role in the AI organization. Artificial Intelligence introduces new considerations related to privacy, intellectual property, regulatory compliance, accountability, and liability. Yet governance should not be viewed as an obstacle to innovation. On the contrary, effective governance enables sustainable innovation by creating clear boundaries within which experimentation can occur responsibly. Organizations that establish appropriate guardrails early are often better positioned to innovate with confidence because they understand the risks they are managing.

Finance provides another critical perspective. In periods of technological excitement, it is easy for organizations to become captivated by possibilities while losing sight of measurable outcomes. Finance leaders bring discipline to the conversation by asking whether investments are generating value. Are productivity levels improving? Are costs declining? Is customer satisfaction increasing? Are risks being reduced? Is decision-making becoming more effective? These questions help ensure that AI remains connected to business performance rather than becoming an exercise in technological enthusiasm. Ultimately, Artificial Intelligence should be evaluated in the same way as any other strategic investment: by the value it creates. One group that deserves particular attention is middle management.

During discussions about organizational transformation, I have often observed that executives and frontline employees receive most of the attention, while middle managers are largely overlooked. This is a mistake. Middle managers occupy a uniquely important position within the organization. They understand operational realities, recognize workflow inefficiencies, appreciate employee concerns, and translate strategic objectives into day-to-day execution. In many cases, they determine whether transformation initiatives succeed or fail. When middle managers are informed, engaged, and empowered, adoption accelerates. When they are excluded from the process, even the most carefully designed strategies can struggle to gain traction.

When all of these groups work together, something significant begins to happen. Artificial Intelligence ceases to be a technology project and becomes an organizational capability. The goal is not to create an AI department operating in isolation from the rest of the enterprise. The goal is to build an AI-enabled organization in which leadership provides strategic direction, data teams manage information effectively, technology teams maintain robust infrastructure,

Human Resources develops talent, legal teams ensure responsible governance, finance measures value, managers drive adoption, and employees embrace continuous learning. At that point, AI becomes far more than software. It becomes part of the organization's operating model. It influences how decisions are made, how opportunities are identified, and how resources are deployed. Most importantly, it becomes a source of competitive advantage that extends beyond the technology itself. Competitors can purchase the same software, license the same platforms, and access many of the same tools. What they cannot easily replicate is an organization that has learned how to align people, processes, governance, and technology around a shared strategic purpose.

That is ultimately the real objective of AI transformation. The long-term advantage does not come from the technology alone. It comes from building an organization that knows how to use technology effectively, responsibly, and in a way that strengthens both performance and human potential. In the years ahead, that capability may prove to be one of the most valuable assets any organization can possess.

Many artificial intelligence initiatives fail for a surprisingly simple reason: no one truly owns them. Responsibility becomes fragmented across the organization. Technology leaders assume AI is a business initiative. Business leaders expect the technology department to take the lead. Human Resources looks to operations, while operations assumes responsibility belongs to information technology. Everyone believes someone else is in charge, and as a result, accountability becomes diluted. Successful organizations avoid this trap by establishing clear ownership from the outset. They recognize that AI is not merely a technology initiative; it is an enterprise-wide transformation that requires leadership, coordination, and sustained commitment.

The Chief Executive Officer does not need to become an expert in machine learning or large language models. That is not the CEO's role. The CEO's responsibility is to champion the transformation, communicate why it matters, and ensure that AI initiatives remain aligned with the organization's strategic objectives.

Visible executive sponsorship is essential. Without it, AI projects often lose momentum as competing priorities emerge and organizational attention shifts elsewhere. Leaders set direction, establish priorities, allocate resources, and reinforce accountability. While they are not responsible for building the technology, they are responsible for leading the organization through the change that technology makes possible.

Once executive commitment has been established, responsibility for coordinating AI efforts must be clearly assigned. Some organizations have created the role of Chief AI Officer, while others distribute those responsibilities to an existing executive. Regardless of the title, someone must own the roadmap, coordinate initiatives across departments, and ensure that individual projects contribute to broader organizational goals rather than becoming isolated experiments. This role has become increasingly important because AI touches nearly every function of the modern enterprise. No department can implement it successfully in isolation.

Data Before Algorithms

As discussed throughout this book, every successful AI initiative begins with data. Sophisticated algorithms cannot compensate for poor-quality, inaccessible, or poorly governed information. This is where the Chief Data Officer, or the executive responsible for data governance, becomes indispensable. That leader ensures that organizational data is accurate, secure, accessible, well managed, and aligned with business priorities.

Many organizations become captivated by the promise of AI while overlooking the condition of their data. Doing so is much like purchasing a sophisticated aircraft before constructing a runway. The technology may be impressive, but without the necessary foundation, it cannot perform as intended.

Building strong data infrastructure is rarely the most visible part of an AI strategy, yet it is often the most important.

Technology as an Enabler

Technology teams continue to play a central role, but their responsibilities extend far beyond deploying software. They evaluate platforms, integrate new systems with existing infrastructure, protect organizational data through strong cybersecurity practices, monitor performance, ensure reliability, and design solutions that can scale as organizational needs evolve.

As AI becomes embedded in daily operations, technology leaders become stewards of an increasingly complex digital ecosystem. Their work creates the stability and resilience that allow innovation to flourish.

Preparing People, Not Just Systems

Perhaps no department will have a greater influence on long-term AI success than Human Resources, or what I prefer to call Humane Resources. The name reflects an important principle. Successful transformation is ultimately about people. One of the most common mistakes organizations make is investing heavily in technology while investing too little in workforce readiness. Employees hear discussions about automation, artificial intelligence, and efficiency. In the absence of clear communication, many naturally begin to wonder what those changes mean for their own future. Uncertainty often creates anxiety, and anxiety frequently leads to resistance. Resistance, in turn, slows organizational transformation. The solution is not simply more technology. It is clearer communication, and thoughtful preparation.

Humane Resources should lead efforts to prepare the existing workforce through skills assessments, actively proactive professional development, reskilling initiatives, leadership training, and career planning. These are strategic investments that help organizations realize the full value of AI while maintaining employee trust and engagement. The organizations that lead in the AI era will not simply deploy intelligent systems. They will prepare their people to work confidently and effectively alongside them.

The Critical Role of Middle Management

One group is often overlooked during AI transformation, despite playing one of the most influential roles in its success: middle management. Organizations frequently focus their attention on executives who establish strategy and frontline employees who use new technologies. Middle managers, however, translate strategy into everyday operations. They understand existing workflows, recognize operational bottlenecks, appreciate employee concerns, and know where practical improvements can be made.

Ignoring middle management often slows adoption because the people responsible for implementation have not been fully engaged. Empowering these leaders, on the other hand, frequently accelerates change by creating advocates who can guide their teams through the transition.

From Technology Initiative to Organizational Capability

When executive leadership, data professionals, technology teams, Humane Resources, legal counsel, finance, operational leaders, and middle managers work together, AI stops being a collection of software applications.

It becomes an organizational capability. The objective is not to create an AI department. The objective is to build an AI-enabled organization in which every function understands its role in creating value.

Leaders provide strategic direction. Data teams ensure information is accurate and trustworthy. Technology professionals build secure and reliable infrastructure. Humane Resources prepares people for change. Legal teams establish governance. Finance measures value. Managers guide implementation, and employees continue developing the skills needed to succeed in an increasingly AI-enabled workplace.

When this alignment occurs, artificial intelligence becomes much more than another technology investment. It becomes a source of sustained competitive advantage. Technology itself can be purchased by almost any organization. What is far more difficult to acquire is a culture that knows how to apply technology thoughtfully, responsibly, and effectively. That is the real objective of AI transformation. It is also where organizations are most likely to achieve their greatest long-term return on investment.

"Humane Resources: Why AI Leadership Is Really About People."

That chapter may end up being the most memorable chapter in the entire book because it is my original contribution to the AI conversation. Most books discuss technology. Very few discuss dignity, workforce preparation, reskilling, and the moral responsibility of leaders in an AI-driven world.

Chapter 23

Humane Resources

Why the Future of AI Must Be About People

As I reflect on everything I have learned through my journey in Artificial Intelligence, from my studies at Oxford to the many executive briefings, boardroom discussions, and professional engagements across different sectors, I find myself returning repeatedly to a single conclusion that becomes clearer the deeper one goes into this field. Artificial Intelligence is not ultimately a technology story. It is a people story.

At first glance, that may appear to be a paradox in a book devoted to one of the most advanced technological shifts of our time. We have explored machine learning, large language models, agentic systems, robotics, automation, and the future of work in considerable detail. These developments matter, and they will continue to shape industries in profound ways. Yet behind every algorithm is a human decision.

Behind every automation initiative is a redesigned role. Teams must learn to adapt far beyond the boundaries of any dashboard or performance metric.

Technology changes systems, processes, and industries, but it also changes lives in ways that are not always immediately visible in boardroom discussions. This is why I believe that leadership in the age of Artificial Intelligence requires a deeper shift in how we think about people within organizations. During my time at Oxford, I often found myself reflecting on a term that appears in almost every organization in the world, Human Resources. It is a familiar phrase, embedded in organizational structures, policies, and governance frameworks. Yet the more I considered the future of work through the lens of AI, the more I began to question whether the language itself was sufficient for the era we are entering.

Human Resources, as a concept, describes people in functional terms. It treats human beings as resources to be managed, allocated, optimized, and deployed. There is nothing inherently wrong with efficiency or organization, and no modern institution can function without structure. However, something subtle but important is lost when people are primarily framed as resources. It becomes easier, often unintentionally, to think of them in terms of output rather than in terms of lives, aspirations, and human complexity.

This is where I have increasingly found value in a different framing, one that emerged from conversations and reflections during my broader AI journey at the University of Oxford. What if, instead of Human Resources, we began to think in terms of Humane Resources?

The difference may appear small at first glance, but the shift in meaning is significant. Humane Resources does not reject structure or performance. Rather, it places human dignity, development, and adaptability at the center of organizational thinking. It acknowledges that people are not simply inputs into a system. They are individuals with histories, ambitions, responsibilities, and futures that extend beyond the workplace.

This distinction becomes particularly important as Artificial Intelligence becomes more capable. For the first time in modern organizational history, we are deploying systems that can perform tasks previously reserved for human effort at scale, speed, and consistency. It is natural, therefore, that many leaders initially approach AI through the lens of efficiency. There are, without question, real efficiencies to be gained. In a competitive environment, organizations that ignore productivity improvements risk falling behind. That reality cannot be dismissed.

Yet efficiency is only one side of the equation. Every efficiency gain carries with it a human dimension that is often less visible but equally important. A model may identify redundancies in a process. A dashboard may highlight opportunities for automation. An algorithm may reveal where tasks can be performed more effectively by machines. But none of these tools can fully capture what those changes mean for the individuals whose roles are affected. A parent supporting a family, a young professional beginning a career, a mid-career employee balancing long-term responsibilities, or someone approaching the end of their working life will each experience transformation differently. These are not abstract data points. They are human realities.

Leadership, therefore, requires the ability to hold both perspectives at once. The economic logic of transformation and the human experience of transformation. The business case and the lived consequence. It is easy to discuss restructuring or automation in conceptual terms, but considerably more difficult when those decisions are reflected in real lives and real futures. This tension, I believe, will become one of the defining leadership challenges of the coming decades.

History offers important context. The Industrial Revolution displaced certain forms of work while creating entirely new industries that were previously unimaginable. The rise of the Internet reshaped communication, commerce, education, and entertainment, while also redefining entire labor markets. Artificial Intelligence will likely follow a similar trajectory. Some roles will disappear, many will evolve, and entirely new categories of work will emerge. The central challenge is not whether change will occur, but how societies and organizations will prepare people for that change.

This is where the responsibility of leadership becomes both strategic and moral. From a strategic perspective, organizations require people who can operate effectively in environments shaped by intelligent systems. From a moral perspective, organizations have a responsibility to ensure that individuals are not left behind by the very systems they help to build and operate. In this context, reskilling cannot be treated as a peripheral initiative or an optional benefit. It becomes a core investment in organizational resilience. Upskilling is not a discretionary activity. It is a structural requirement for long-term competitiveness. Continuous learning is not an aspirational value. It is a necessary condition for survival in an environment of constant technological change.

The organizations that succeed in the long term may not necessarily be those with the most advanced AI systems. More likely, they will be those that develop an exceptional capacity to help people adapt alongside technology. In many respects, the defining capability of the future workforce will not be static expertise but sustained adaptability. The ability to learn, unlearn, and relearn will matter as much as any specific technical skill. Those who thrive will be individuals who remain curious, open to change, and willing to reinvent themselves when circumstances require it.

At the same time, it is important to recognize that Artificial Intelligence, for all its power, does not diminish the importance of what makes us fundamentally human. Empathy, judgment, creativity, integrity, and wisdom remain beyond the reach of any algorithm. These are not technical skills that can be automated or replicated through computation. In fact, as technology becomes more capable, these human qualities may become more important, not less. Leadership, in particular, will continue to depend on the ability to interpret complexity, navigate uncertainty, and make decisions that balance performance with principle.

This is why I remain optimistic about the future, although not in a simplistic or uncritical sense. I do not believe Artificial Intelligence will solve every problem, nor do I believe that its adoption will be smooth or evenly distributed. Disruption is inevitable, and transition will be uneven across industries and regions. My optimism comes instead from a longer view of human history, which consistently demonstrates that societies are capable of adapting to profound technological change. Time and again, human beings have shown an extraordinary ability to innovate, adjust, and ultimately find new forms of progress in the face of disruption.

The question, therefore, is not whether Artificial Intelligence will transform the future of work. It will. The question is whether that transformation will diminish human potential or expand it. Whether it will narrow opportunity or broaden it. Whether it will create exclusion or enable inclusion. That outcome is not predetermined by technology. It is shaped by decisions made in leadership teams, boardrooms, governments, and institutions around the world.

Technology can create capability, efficiency, and scale, but it cannot determine direction. Only leadership can do that. And leadership, at its core, is a human responsibility. It requires judgment, empathy, and a willingness to think beyond short-term optimization in favor of long-term impact.

This is why, when I think about the future of Artificial Intelligence, I do not primarily think about machines. I think about people. Because ultimately, the success of AI will not be measured by how intelligent our systems become, but by whether human lives are improved, expanded, and made more meaningful as a result of their existence. That is the standard we should hold ourselves to. And that is the responsibility we collectively share. If there is one final lesson that history consistently teaches us about Artificial Intelligence, it is humility in prediction. The pioneers of earlier eras underestimated both the difficulty and the trajectory of progress. Even in recent years, technologies have arrived faster than many expected, while others have evolved in ways few predicted with accuracy. The future, therefore, should not be approached with certainty, but with responsibility, awareness, and a commitment to ensure that whatever emerges next remains grounded in human purpose.

If there is one lesson I have learned throughout my journey into Artificial Intelligence, it is that no one truly knows what comes next. Not the professors who study it, the researchers who advance it, the technology companies that build it, the governments attempting to regulate it, the investors funding its growth, or even many of the engineers developing the systems themselves. We can make informed predictions, identify emerging trends, and analyze patterns from the past, but history has repeatedly shown that technological revolutions rarely unfold exactly as expected. Innovation has a way of surprising even the people closest to it.

When I began my studies at Oxford in 2023, Artificial Intelligence was already becoming one of the defining technological conversations of our time. Researchers and business leaders understood its potential, but relatively few anticipated the extraordinary speed at which the field would evolve over the following years. Capabilities that had once seemed several years away began appearing within months. Sophisticated tools that previously required specialized expertise became accessible to the public, while tasks that once consumed hours of manual effort could often be completed in minutes with the assistance of AI. The pace of progress was remarkable, and I believe we are still witnessing the earliest stages of a much larger transformation.

Among the developments that I find most significant is the emergence of Agentic AI. Most AI systems today are designed to respond to individual requests by generating text, analyzing information, or completing specific tasks. Increasingly, however, developers are building systems capable of planning, coordinating, and carrying out sequences of interconnected activities with limited human supervision. Rather than simply drafting a report, future AI agents may gather information from multiple sources, analyze data, prepare recommendations, identify potential risks, organize meetings, coordinate stakeholders, monitor project milestones, and support implementation from beginning to end. If that evolution continues, the role of Artificial Intelligence may gradually shift from being a useful assistant to becoming a collaborative partner in many professional environments.

At the same time, AI technologies are likely to become far more integrated than they are today. Most organizations currently rely on numerous specialized applications, using one platform for writing, another for image generation, another for video creation, another for data analysis, and still others for project management or workflow automation. Although these tools are increasingly powerful, the experience often remains fragmented because users must move continually between separate systems. Over time, that fragmentation may diminish as AI platforms combine multiple capabilities into unified environments capable of executing complex objectives through a single instruction.

Consider a nonprofit organization seeking to expand its educational programs into three new regions. Today, accomplishing that objective typically requires extensive coordination among researchers, financial analysts, designers, project managers, and communications teams. In the future, an integrated AI system may assist throughout much of that process by gathering demographic research, evaluating financial assumptions, preparing strategic recommendations, drafting presentations, creating supporting visuals, organizing implementation timelines, and scheduling follow-up actions. While elements of this workflow already exist in limited forms, continued advances in AI suggest that these capabilities may become increasingly seamless and interconnected.

The transformation extends beyond software alone. Robotics continues to advance rapidly, expanding well beyond its traditional role in manufacturing. Robots are now being deployed in healthcare, planning, agriculture, construction, warehousing, and disaster response, where they perform tasks that may improve efficiency, safety, or precision. Advances in drone technology, satellite imaging, autonomous systems, and intelligent sensors are also reshaping industries by providing richer data and enabling more informed decisions based on real-time information. As these technologies mature and become more affordable, they are likely to influence not only how organizations operate but also how societies address challenges in areas ranging from infrastructure and environmental management to healthcare and emergency. Some of the most important questions remain profoundly human. How should these technologies be governed? How can societies encourage innovation while maintaining accountability? How do we protect privacy without limiting beneficial uses of data? How should educational institutions prepare students for occupations that may look very different within a decade?

And how can governments, businesses, and communities work together to ensure that the benefits of Artificial Intelligence are shared broadly rather than concentrated among a relatively small number of organizations or individuals? These are not questions that algorithms can answer on their own. They require thoughtful leadership, sound judgment, ethical reflection, and a commitment to ensuring that technological progress ultimately serves the public good.

Another subject that frequently dominates discussions about AI is Artificial General Intelligence, commonly referred to as AGI. Although definitions vary, AGI generally describes a hypothetical system capable of performing intellectual tasks across a wide range of domains at a level comparable to or potentially exceeding—that of humans. Whether such systems will ultimately be achieved remains uncertain. Some researchers believe AGI could emerge within the coming years, while others argue that it may require decades of additional scientific progress or question whether it is attainable at all. At present, no one can answer that question with confidence.

What is clear is that today's AI systems, despite their remarkable capabilities, remain specialized tools with important limitations. Large language models, including the system that helped refine portions of this manuscript, are trained on vast collections of information, but they do not possess continuous awareness of everything happening in the world. Their responses depend on the data available during training together with any external tools or information sources to which they have access. Appreciating both the strengths and the limitations of these systems is essential for using them responsibly and effectively.

One certainty is that Artificial Intelligence will continue to evolve. Today's breakthroughs will become the foundation upon which future innovations are built. New architectures will emerge, new capabilities will be developed, and new applications—many of which we cannot yet imagine—will reshape industries, institutions, and everyday life. At the same time, each new advance will introduce fresh challenges requiring careful thought, responsible governance, and continued adaptation.

That reality is one reason I doubt this will be the last book I write about Artificial Intelligence. The field is simply moving too quickly for any single volume to remain definitive for long. By the time these pages reach readers, new companies will have emerged, new tools will have been introduced, and new discoveries will already be reshaping what AI can accomplish. Unlike many subjects that can be documented as completed histories, Artificial Intelligence is a story that continues to unfold in real time.

Perhaps that is what makes this moment so remarkable. We are not reading the concluding chapter of Artificial Intelligence; we are witnessing the opening chapters of one of the most consequential technological transformations in modern history. The future has not yet been determined, and the decisions we make today will help shape how these technologies develop and how they influence future generations. Whether we serve as educators, business leaders, entrepreneurs, policymakers, researchers, parents, or engaged citizens, each of us has a role to play in guiding the responsible development and application of Artificial Intelligence.

The question is not whether Artificial Intelligence will shape the future. It almost certainly will. The more important question is whether we will exercise the wisdom, responsibility, and foresight necessary to shape Artificial Intelligence in ways that expand opportunity, strengthen society, and improve the human condition. I believe that responsibility belongs to all of us, and I remain optimistic that it is one we are capable of fulfilling.

Epilogue

The Promise

When I began my studies at Oxford, I made a quiet promise to myself that I would write a book about Artificial Intelligence at Graduation. At the time, I could not have known what that promise would eventually demand of me, nor could I have anticipated how quickly the field itself would evolve while I was still learning it. I did not know how many conversations I would have, how many leaders I would meet, or how many organizations I would observe as they wrestled with the same fundamental questions of opportunity, uncertainty, and transformation. What I did know, however, was that I was witnessing something historically significant, something that would reshape not only industries but also the way we think about work, knowledge, and decision-making itself.

As my time at Oxford unfolded, Artificial Intelligence gradually ceased to be an abstract subject of study. It became a lens through which I began to interpret almost everything around me, from business strategy and economic policy to education systems and the evolving nature of human potential. It challenged long-held assumptions, expanded what I believed to be possible, and at times forced me to reconsider ideas I had previously taken for granted. Yet when I look back now, what remains most vivid in my memory is not the technical material or the academic structure of the program. It is the people.

It is the professors who shared their insights with generosity and intellectual curiosity, the classmates who arrived from different countries, each bringing with them a different perspective shaped by their own industries and experiences. It is the conversations that continued long after formal sessions ended, often becoming as meaningful as the lectures themselves. We developed a sense of shared discovery learning alongside professionals equally committed to understanding a rapidly changing world.

Among those who shaped this journey most deeply was my twin sister. She believed in my decision to study at Oxford in a way that was unwavering and deeply personal. She celebrated each milestone with genuine joy and encouraged me through moments of doubt with a quiet strength that I often drew upon more than she probably realized. She was one of my greatest supporters, and I had always imagined her presence at my graduation as part of that journey coming full circle.

Life, however, had other plans. Before I completed my studies, she completed her own life journey. Even now, writing those words carries a weight that is difficult to fully express. She had worked tirelessly in pursuit of her own educational goals, persevering through challenges that required both courage and resilience.

She completed her examinations, but she was not here to see the results of her efforts, nor to stand beside me at a moment we had both quietly looked forward to. There is a sadness in that reality that does not diminish with time. Yet alongside that sadness is a profound sense of gratitude. Gratitude for her presence in my life. Gratitude for her belief in me, and gratitude for the shared years that shaped so much of who I am today.

As I approach graduation, I find myself reflecting more deeply on something that no technological system, no matter how advanced, could ever truly replicate. The essence of what makes us human. It cannot replace love. It cannot replace friendship. It cannot replace sacrifice. It cannot replace faith. And it cannot replace the meaning we attach to the relationships that shape our lives.

It is in this realization that I find one of the most important lessons of my entire journey in this second chapter of my life and career, after my twin and after oxford. As the future becomes more technologically advanced, it must also remain deeply human to survive. Progress that is measured only in computational power or efficiency is incomplete. True progress must also be measured in empathy, in human dignity, in connection, and in the quality of the lives we lead alongside the systems we create.

As I look ahead, I remain optimistic because of an enduring belief in people and our ability to adapt, and overcome extreme challenges. If we approach this technology with wisdom, humility, curiosity, and compassion, then it is possible that we will not only become more productive, but also more thoughtful in how we define progress itself. The future, in that sense, would not be predetermined by machines. It will be shaped by the choices we make about how to use them, and more importantly, by what we choose to preserve about ourselves in the process.

This book began with a journey to Oxford, but it has ended as something larger than I expected. It has become a reflection on learning, leadership, and what it means to remain human in a time of accelerating technological change. In many ways, both the journey and the questions it raises are still unfolding.

One of the most prevailing lessons I carry forward is that the most important dimensions of life often cannot be measured in data or captured in models. There are experiences that exist beyond analysis, moments that resist explanation, and connections that remain meaningful precisely because they cannot be reduced to logic alone. My sister's memory is part of that understanding. Her presence in my life, and the impact she continues to have on my thinking, reminds me that some of the most important truths we live with are not derived from computation but from lived experience, memory, and love.

As I prepare to walk across the stage at Oxford, I do so with a sense of gratitude that feels both personal and expansive. Gratitude for the opportunity to learn and to the people who shaped this journey. From my loving husband, to my entire family, friends, admin staff, and professors at Oxford.

I have Gratitude for the responsibility of living in a time when technology holds such extraordinary potential, and above all, for the reminder that even in a world increasingly shaped by Artificial Intelligence, it is still human intelligence, human emotion, and human connection that provide that evolution with meaning.

If there is one thought I hope remains with the reader beyond this book, it is a simple one. As we build more intelligent machines, we must also strive to become more thoughtful, more compassionate, and more aware of one another. Because in the end, the true measure of any technological era is not how advanced its tools become, but whether those tools help us become better versions of ourselves.

That is the future I hope we choose to build.

Lagniappe

I could not have anticipated how rapidly this technology would evolve, how many organizations I would have the privilege of speaking with, or how many conversations I would have with leaders seeking to understand what AI might mean for their businesses, their professions, and their futures. I simply knew that I was witnessing the beginning of something historically significant, and I wanted to document that moment with honesty, curiosity, and optimism.

What began as an academic pursuit gradually became something much larger. Artificial Intelligence was no longer just a subject I studied, but a lens through which I encountered and examined leadership, innovation, deeper education, public policy, and the changing nature of work. The technology challenged long-held assumptions, expanded my thinking, and continually reminded me that periods of rapid change demand both intellectual curiosity and humility.

Yet when I reflect on my time at Oxford, the memories that remain most vivid are not the lectures, the research, or even the technology itself. They are the people. I remember professors whose generosity with their knowledge inspired every student they encountered. I remember classmates who traveled from every corner of the world, bringing with them different cultures, experiences, and perspectives that enriched every conversation. Some of the most meaningful learning happened long after class had ended, during discussions that continued over coffee, dinners, and countless conversations about the future we were all trying to understand together.

Most of all, I remember those who encouraged me long before this book was finished. Among them was my twin sister, **Uju Benedicta**

As my graduation approaches, I find myself reflecting on a coincidence that I cannot easily dismiss. My sister passed away on January 11, and my Oxford graduation is scheduled for July 11, sharing the same day of the month. She also passed away at 56. 5+6 = 11. Without planning or requesting it, her grave number was 56. The sister who helped care for us during our childhood happened to live in Apartment 56. The hotel room we were assigned when we checked into on the day of her funeral, was 56.

Some might say perhaps it is simply coincidence or that it reflects the natural tendency of the human mind to search for meaning during seasons of both grief and celebration. I do know one thing and that is that there is a higher power behind our universe. This is also one distinction between us and AI. My book, God Gives us Miracles available on Amazon.com is full of miracle stories from real people.

Some people may see nothing more than coincidence in these details, while others may see something deeper. What matters most is acknowledging that experiences like these remind us that there are dimensions of life that cannot be fully explained through data alone.

Throughout this book, I have written about algorithms, models, automation, and the extraordinary capabilities of Artificial Intelligence. These systems can process enormous amounts of information, recognize complex patterns, generate sophisticated content, and assist with solving problems that once seemed beyond our reach. They are remarkable achievements of human ingenuity.

Yet for all their power, there are aspects of the human experience that no technology can fully reproduce.

Artificial Intelligence can identify patterns, but it cannot experience love. It can generate language, but cannot feel compassion. It can analyze grief without ever mourning a loss, describe hope without ever needing it, and explain faith without ever possessing it. It can help us make decisions, but it cannot determine what gives those decisions meaning. Those qualities remain uniquely human.

That realization has become, for me, the most important lesson of this entire journey. As our technologies become more capable, our responsibility to preserve the qualities that define our humanity becomes even greater. Intelligence alone is not enough. The future will also require wisdom, integrity, compassion, humility, and a deep commitment to using innovation in ways that strengthen rather than diminish the human experience.

I remain profoundly optimistic about what lies ahead. Not because I believe Artificial

Intelligence will solve every problem, but because I continue to believe in people. I believe in our ability to lead responsibly, and to build technologies that reflect our highest values rather than our lowest instincts. If we approach this era with wisdom, curiosity, and compassion, I believe we can create a future that is not only more productive but also more just, more thoughtful, and more humane.

My Oxford journey began with a promise to write this book. Completing these final pages feels like fulfilling that promise and I'm sure my twin is smiling with her "AI Chief", as she used to call me.

At the same time, I know this is not an ending. Artificial Intelligence will continue to evolve. New discoveries will reshape industries, new challenges will emerge, and new questions will demand thoughtful answers. This book captures one moment in that unfolding story, but the story itself is only beginning. I look forward to Volume 2 and seeing you around.

For readers interested in the more personal experiences that shape my perspective, I explore many of those events in my companion book, *God Gives Us Miracles,* available on Amazon.com. While this AI book examines the future of technology, the one on Miracles reflects on the experiences of faith, family, and God's Hands in our lives.

As I prepare to walk across the stage at Oxford to receive my diploma, I do so with profound gratitude to God, for the education I received, the mentors and friends who enriched the journey, the opportunities that made it possible, and the people whose love sustain me through every season.

Above all, I am grateful to be living at a time in history when humanity has the extraordinary opportunity not only to witness one of by far, the greatest technological transformations ever experienced, but also to help shape its direction. The choices we make today will influence not only the future of Artificial Intelligence, but the future of our communities, our institutions, and the generations that follow.

My hope is that we embrace this responsibility with wisdom, humility, and compassion. May we never allow our pursuit of increasingly intelligent machines to overshadow the enduring value of human dignity, the beauty of our shared humanity, or the responsibility we bear, to care for the remarkable planet we all call home.

Technology should not simply make us more efficient, it should help us become better stewards of one another and of the world entrusted to us. If our innovations leave humanity better, more compassionate, more just, and our planet healthier than we found it, then our greatest achievement will not be the intelligence we created, but the legacy we have chosen to leave behind.

References

Ashish Vaswani, N. S. (2017). Attention is all you need. Advances in Neural Information Processing Systems 30.

Center, O. (n.d.). What are tokens and how to count them?

Hsu, F. (2002). *Behind Deep Blue's: Buildingthe computer that defeated the world chess champion.* Princeton University Press.

OpenAI. (.2023). *GPT-4 Technical Report.* OpenAI.

OpenAI. (2022). *Introducing ChatGPT.* OpenAI.

OpenAI. (n.d.). What are tokens and how to count them?

OpenAI. (n.d.). *What are tokens and how to count them?*

technology, N. I. (2024). *NIST AI 600-1: Artificial Intelligence Risk Management Framework: Generative Artificial Intelligence Profile.*

Turing, A. (1950). *Computing machinery and intelligence. Mind, 59(236). 433-460. HTTPS://doi.org/10.1093/ mind/dot 236.433.*

Vaswani, A. S. (2017). Attention is all you need. Advances in Neural Information Processing Systems.

www.ingramcontent.com/pod-product-compliance
Lightning Source LLC
LaVergne TN
LVHW022156200926
848481LV00006B/1766
9781735766560